Country HOME PLANS

All new 3rd edition

Publisher: **James D. McNair III**
Chief Operating Officer: **Bradford J. Kidney**
Staff Writers: **Debra Cochran/Sue Barile**
Cover Design & Interior Layouts: **Paula Mennone**
Back Cover Design: **Josephine Panno**
Illustrations: **Robert Miles Long**

Submit all Canadian plan orders to:
The Garlinghouse Company
60 Baffin Place, Unit #5
Waterloo, Ontario N2V 1Z7

Canadian Orders Only: 1-800-561-4169
Fax No. 1-800-719-3291
Customer Service No.: 1-519-746-4169

Library of Congress No.: 97-77622
ISBN: 0-938708-79-1

PLEASE NOTE: Sunflower design adapted from sunflower wallcovering border in the SUNFLOWERS collection by Sunworthy Wallcoverings. The border is PKB 206. Sunworthy wallcoverings are available nationwide. Information on dealers is available by calling Sunworthy at 800-425-7336.

PHOTOGRAPHY BY JOHN EHRENCLOU

Simply the Best

FIRST FL.	1,622 sq. ft.
SECOND FL.	1,156 sq. ft.
BEDROOMS	Three
BATHROOMS	2(Full), 1(Half)
FOUNDATION	Basement
TOTAL LIVING AREA	**2,778 sq. ft.**

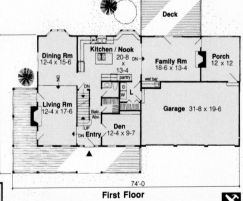

First Floor

Wide corner boards, clapboard siding, and a full-length covered porch lend a friendly air to this classic home with a Colonial accent. The central entry opens to a cozy den on the right, a sunken, fireplaced living room with adjoining dining room on the left, and straight past the powder room to a gourmet's dream kitchen. Bay windows in the informal dining nook lend a cheerful atmosphere to the entire kitchen area, accentuated by atrium doors linking the adjoining, sunken family room with both the three-season porch and rear deck. The two front bedrooms upstairs are served by a large, double-vanitied bath. The master suite spans the rear of the house, enjoying a huge, walk-in closet, a private bath with double vanities, a raised whirlpool tub, and step-in shower. The photographed home has been modified to suit individual tastes.

Second Floor

Refer To Price Code E on Page 253

DESIGN 10805

W9-DAY-149

Southern Belle

The wrap-around porch on this home adds to the classic styling and gives curb appeal. Upon entering the home, the formal living room and dining room are to your left and right, respectively. Columns accentuate the Great room, while the fireplace adds to the mood of the room. A well-appointed kitchen is conveniently placed between the formal dining room and the informal breakfast room. There is a screened porch and a sun deck for outdoor living space. Four bedrooms occupy the second floor. The master suite includes a private master bath, walk-in closet and a private deck. The three additional bedrooms share the full hall bath. The photographed home has been modified to suit individual tastes.

DESIGN 932

ESIGN 93209

PHOTOGRAPHY BY JOHN EHRENCLOU

PLAN INFO

FIRST FL.	1,250 sq. ft.
SECOND FL.	1,166 sq. ft.
FINISHED STAIRCASE	48 sq. ft.
BASEMENT	448 sq. ft.
GARAGE	706 sq. ft.
BEDROOMS	Four
BATHROOMS	2(Full), 1(Half)
FOUNDATION	Basement
TOTAL LIVING AREA	**2,464 sq. ft.**

Refer To Price Code D on Page 253

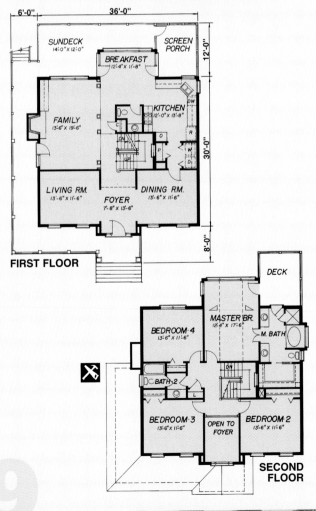

FIRST FLOOR

6'-0"   36'-0"

SUNDECK
14'-0" X 12'-0"

SCREEN PORCH

BREAKFAST
12'-4" X 11'-8"

KITCHEN
12'-0" X 13'-8"

FAMILY
13'-6" X 19'-6"

12'-0"

30'-0"

LIVING RM.
13'-6" X 11'-6"

FOYER
7'-8" X 13'-6"

DINING RM.
13'-6" X 11'-6"

8'-0"

SECOND FLOOR

DECK

BEDROOM 4
13'-6" X 11'-6"

MASTER BR.
12'-4" X 17'-6"

M. BATH

BATH 2

BEDROOM 3
13'-6" X 11'-6"

OPEN TO FOYER

BEDROOM 2
13'-6" X 11'-6"

An
EXCLUSIVE DESIGN
By Karl Kreeger

9

Captivating Charm

Picture a porch swing, cozy rocking chairs and a pitcher of lemonade on this country porch. What an inviting picture. The homey feel continues throughout this house. The formal areas are located in the traditional places, flanking the entry hall. The living room includes a wonderful fireplace and the dining room has direct access to the kitchen. The U-shaped kitchen includes a breakfast bar, built-in pantry, planning desk and a double sink. A mudroom entry will help keep the dirt from muddy shoes away from the rest of the house. A convenient laundry area is close at hand in the half-bath off the mudroom. The sunny breakfast nook is a cheerful place to start your day, and the expansive family room has direct access to the rear wood deck. Sleeping quarters are located on the second floor. The master suite is highlighted by a walk-in closet and private master bath. The two additional bedrooms, one with a built-in desk, share a full hall bath with a double vanity. A window seat in the hallway provides a cozy place to curl up with a book. In fact, bookshelves have been built-in on either side of the seat. The photographed home has been modified to suit individual tastes.

DESIGN 242

PHOTOGRAPHY BY JOHN EHRENCLOU

PLAN INFO

FIRST FL.	1,113 sq. ft.
SECOND FL.	970 sq. ft.
BASEMENT	1,113 sq. ft.
GARAGE	480 sq. ft.
BEDROOMS	Three
BATHROOMS	2(Full), 1(Half)
FOUNDATION	Basement, Slab, Crawl Space
TOTAL LIVING AREA	**2,083 sq. ft.**

Refer To Price Code C on Page 253

Crawl Space/Slab Option

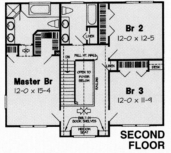

Br 2
12-0 x 12-5

Master Br
12-0 x 15-4

OPEN TO FOYER BELOW

Br 3
12-0 x 11-9

BUILT-IN BOOK SHELVES

WINDOW SEAT

SECOND FLOOR

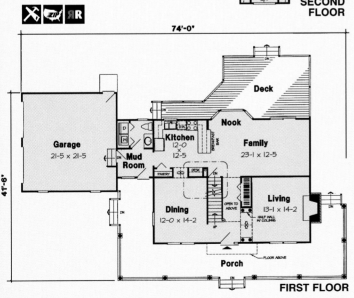

74'-0"

41'-6"

Garage
21-5 x 21-5

Mud Room

Deck

Nook

Kitchen
12-0 x 12-5

Family
23-1 x 12-5

Dining
12-0 x 14-2

Living
13-1 x 14-2

OPEN TO ABOVE

PANTRY

Porch

FLOOR ABOVE

FIRST FLOOR

Country Appeal

T his home combines the best of all worlds: a classic Victorian exterior and an interior plan that more than answers the demands of modern family life. Step past the covered porch into the tiled foyer dominated by an open staircase. A short hall leads past the formal dining room to the first-floor master suite. To the right, the angular kitchen, sunny breakfast bay, and fireplaced hearth room flow together into a pleasing, open space that's perfect for family gatherings. When guests arrive, show them into the spectacular skylit atmosphere of the soaring living room with its adjoining outdoor deck. For a better view, walk upstairs to the balcony that links three bedrooms, two full baths, and a laundry room. The photographed home has been modified to suit individual tastes.

DESIGN 2017

PHOTOGRAPHY BY JOHN EHRENCLOU

PLAN INFO

FIRST FL.	1,625 sq. ft.
SECOND FL.	916 sq. ft.
BASEMENT	1,618 sq. ft.
GARAGE	521 sq. ft.
BEDROOMS	Four
BATHROOMS	3(Full), 1(Half)
FOUNDATION	Basement
TOTAL LIVING AREA	**2,541 sq. ft.**

Refer To Price Code D on Page 253

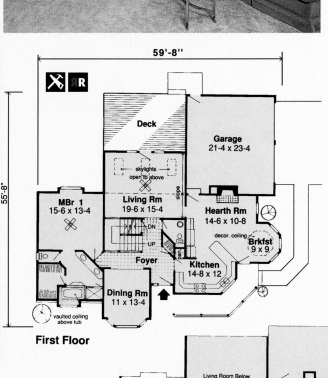

Deck

Garage
21-4 x 23-4

skylights
open to above

slope

MBr 1
15-6 x 13-4

Living Rm
19-6 x 15-4

Hearth Rm
14-6 x 10-8

decor. ceiling

Brkfst
9 x 9

DN
UP

pan.

Foyer

Kitchen
14-8 x 12

Dining Rm
11 x 13-4

vaulted ceiling
above tub

59'-8''

55'-8"

First Floor

An
EXCLUSIVE DESIGN
By Karl Kreeger

Living Room Below

D   W

Br 4
11 x 10-4

DN
Balcony

lin.

Br 2
14-8 x 13-8

Br 3
11 x 11

Second Floor

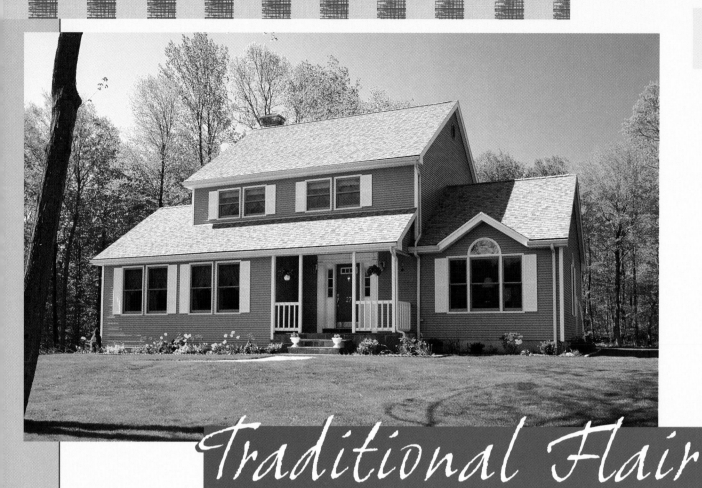

Traditional Flair

Enter into the spacious foyer, walk past the powder room and sloped-ceilinged living/dining room arrangement and into the fire-placed family room. Look to your right and view the modern kitchen with plenty of counter space and a breakfast bar. Upstairs, you'll find two large bedrooms and a master suite complete with a tub, a step-in shower and double vanities in the master bath. For more than ample storage space, use the spacious, two-car garage. The photographed home has been modified to suit individual tastes.

DESIGN 3487

PHOTOGRAPHY BY JOHN EHRENCLOU

PLAN INFO

FIRST FL.	1,088 sq. ft.
SECOND FL.	750 sq. ft.
BASEMENT	750 sq. ft.
GARAGE	517 sq. ft.
BEDROOMS	Three
BATHROOMS	2(Full), 1(Half)
FOUNDATION	Basement, Slab, Crawl Space
TOTAL LIVING AREA	**1,838 sq. ft.**

Refer To Price Code C on Page 253

SECOND FLOOR

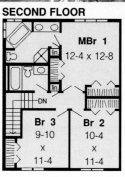

MBr 1
12-4 x 12-8

DN

Br 3
9-10
x
11-4

Br 2
10-4
x
11-4

D W

Slab/Crawlspace
Option

optional
Deck

Dining
12-6 x 10-6
slope

Kitchen
14-9 x 8-0

Family Rm
21-3 x 13-8

optional
Fireplace

wood
storage

W D

Living Rm
12-6 x 14-6
slope

DN

Foyer

UP

Garage
23-6 x 23-4

36'-8"

50'-0"

FIRST FLOOR

Fabulous Facade

You'll never get bored with the rooms in this charming, three-bedroom Victorian. The angular plan gives every room an interesting shape. From the wrap-around veranda, the entry foyer leads through the living room and parlor, breaking them up without confining them, and giving each room an airy atmosphere. In the dining room, with its hexagonal recessed ceiling, you can enjoy your after-dinner coffee and watch the kids playing on the deck. Or eat in the sunny breakfast room off the island kitchen, where every wall has a window, and every window has a different view. You'll love the master suite's bump-out windows, walk-in closets, and double sinks. The photographed home has been modified to suit individual tastes.

DESIGN 3492

PHOTOGRAPHY BY JOHN EHRENCLOU

PLAN INFO

FIRST FL.	1,409 sq. ft.
SECOND FL.	1,116 sq. ft.
BASEMENT	1,409 sq. ft.
GARAGE	483 sq. ft.
BEDROOMS	Three
BATHROOMS	2(Full), 1(Half)
FOUNDATION	Basement, Slab, Crawl Space
TOTAL LIVING AREA	**2,525 sq. ft.**

Refer To Price Code D on Page 253

BEDROOM 3
12'-4"
X
13'-4"

BEDROOM 2
12'-4"
X
13'-4"

M.BEDROOM
14'-0"
X
15'-4"

SECOND FLOOR

opt. slab/ crawl space

An
EXCLUSIVE DESIGN
By Karl Kreeger

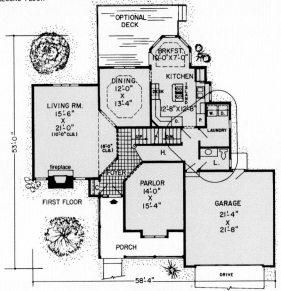

OPTIONAL
DECK

BRKFST.
10'-0"X7'-0"

DINING
12'-0"
X
13'-4"

KITCHEN
12'-8"X12'-8"

LIVING RM.
15'-6"
X
21'-0"
(10'-0" CLG.)

LAUNDRY

(8'-0"
CLG.)

fireplace

FOYER

FIRST FLOOR

PARLOR
14'-0"
X
15'-4"

GARAGE
21'-4"
X
21'-8"

PORCH

53'-0"

58'-4"

DRIVE

The Epitome of Country

Victorian elegance combines with a modern floor plan to make this a dream house without equal. A wrap-around porch and rear deck add lots of extra living space to the roomy first floor, which features a formal parlor and dining room just off the central entry. Informal areas at the rear of the house are wide-open for family interaction. Gather the crew around the fireplace in the family room, or make supper in the kitchen while you supervise the kids' homework in the sun-washed breakfast room. Three bedrooms, tucked upstairs for a quiet atmosphere, feature skylit baths. And, you'll love the five-sided sitting nook in your master suite, a perfect spot to relax after a luxurious bath in the sunken tub. The photographed home has been modified to suit individual tastes.

DESIGN 1069

PHOTOGRAPHY BY JOHN EHRENCLOU

PLAN INFO

FIRST FL.	1,260 sq. ft.
SECOND FL.	1,021 sq. ft.
BASEMENT	1,186 sq. ft.
GARAGE	851 sq. ft.
BEDROOMS	Three
BATHROOMS	2(Full), 1(Half)
FOUNDATION	Basement, Slab, Crawl Space
TOTAL LIVING AREA	**2,281 sq. ft.**

Refer To Price Code D on Page 253

Alternate Crawl/Slab Plan

Br #3
11-7 x 9-10

MBr #1
12-1 x 15-10
8' Clg.

Br #2
11-7 x 11-10

Open to Below

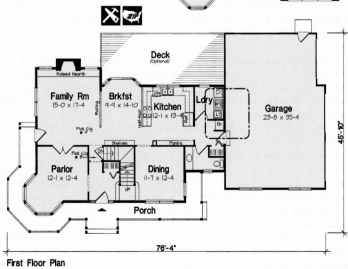

Deck
(Optional)

Family Rm
15-0 x 17-4

Brkfst
9-9 x 14-10

Kitchen
12-1 x 13-4

Ldry

Garage
23-8 x 35-4

45'-10"

Parlor
12-1 x 12-4

Dining
11-7 x 12-4

Porch

76'-4"

First Floor Plan

A Touch of Class

Does your family enjoy entertaining? Here's your home! This handsome, rambling beauty can handle a crowd of any size. Greet your guests in a beautiful foyer that opens to the cozy, bayed living room and elegant dining room with floor-to-ceiling windows. Show them the impressive two-story gallery and book-lined study, flooded with sunlight from atrium doors and clerestory windows. Or, gather around the fire in the vaulted family room. The bar connects to the efficient kitchen, just steps away from both nook and formal dining room. And, when the guests go home, you'll appreciate your luxurious first-floor master suite and the cozy upstairs bedroom suites with adjoining sitting room. The photographed home has been modified to suit individual tastes.

DESIGN 1066

PHOTOGRAPHY BY JOHN EHRENCLOU

PLAN INFO

FIRST FL.	2,310 sq. ft.
SECOND FL.	866 sq. ft.
GARAGE	679 sq. ft.
BEDROOMS	Three
BATHROOMS	3(Full), 1(Half)
FOUNDATION	Slab
TOTAL LIVING AREA	**3,176 sq. ft.**

Refer To Price Code E on Page 253

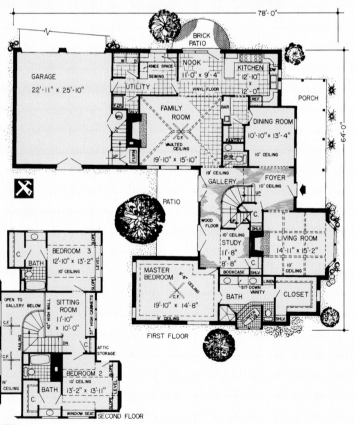

FIRST FLOOR

SECOND FLOOR

PHOTOGRAPHY BY JON RILEY OF RILEY & RILEY PHOTOGRAPHY

Special Touches

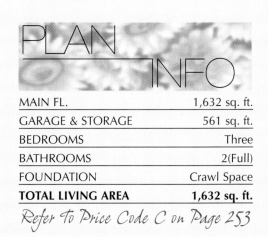

This country home is as beautiful from the back as it is from the front. Porches front and back, gables, and dormers provide special charm. The central Great room has a cathedral ceiling, fireplace, and a clerestory window which brings in lots of natural light.

Columns divide the open Great room from the kitchen and breakfast bay. A tray ceiling and columns dress up the formal dining room. The master suite with tray ceiling and back porch access is privately-located in the rear. The skylit master bath features whirlpool tub, shower, dual vanity, and spacious walk-in closet. The front bedroom with walk-in closet doubles as a study. A garage with ample storage completes the plan. The photographed home has been modified to suit individual tastes.

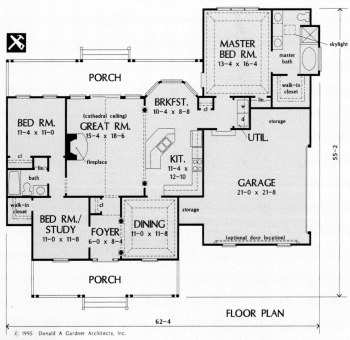

FLOOR PLAN

PORCH

BED RM.
11-4 x 11-0

(cathedral ceiling)
GREAT RM.
15-4 x 18-6

fireplace

BRKFST.
10-4 x 8-8

MASTER
BED RM.
13-4 x 16-4

master bath

skylight

walk-in closet

lin.

storage

w d

UTIL.

KIT.
11-4 x 12-10

GARAGE
21-0 x 21-8

walk-in closet

bath

lin.

cl

BED RM./
STUDY
11-0 x 11-8

FOYER
6-0 x 8-4

DINING
11-0 x 11-8

storage

(optional door location)

PORCH

62-4

55-2

© 1995 Donald A Gardner Architects, Inc.

PLAN INFO

MAIN FL.	1,632 sq. ft.
GARAGE & STORAGE	561 sq. ft.
BEDROOMS	Three
BATHROOMS	2(Full)
FOUNDATION	Crawl Space
TOTAL LIVING AREA	**1,632 sq. ft.**

Refer To Price Code C on Page 253

DESIGN 99840

Country Pleasures

This beautiful home accommodates the needs of a growing family and looks stunning in any neighborhood. The porch serves as a wonderful relaxing area to enjoy the outdoors. At the rear of the home is a patio, for a barbeque or for private time away from the kids. Inside is just as delightful. The dining room features a decorative ceiling and has a convenient entry to the kitchen. The kitchen/utility area has a side exit into the garage. The living room has double doors into the fireplaced family room which features a back entrance to the patio. Upstairs is the sleeping area with three bedrooms plus a vaulted-ceiling master bedroom. The master bedroom has two enormous walk-in closets, as well as a dressing area and private bath. The photographed home has been modified to suit individual tastes.

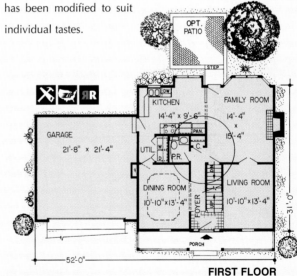

FIRST FLOOR

PLAN INFO

FIRST FL.	955 sq. ft.
SECOND FL.	1,005 sq. ft.
BASEMENT	930 sq. ft.
GARAGE	484 sq. ft.
BEDROOMS	Four
BATHROOMS	2(Full), 1(Half)
FOUNDATION	Basement, Slab, Crawl Space
TOTAL LIVING AREA	**1,960 sq. ft.**

Refer To Price Code C on Page 253

SECOND FLOOR

Slab/Crawlspace Option

An **EXCLUSIVE DESIGN**
By Karl Kreeger

DESIGN 34027

Endless Possibilities

A house is not a home until the family gives it heart. This plan is ready for your family to give a heart. The formal areas are place at the front of the home with a view of the wrap-around porch and the front yard. The informal, family, areas are located in the center and rear of the home. The family room incorporates a fireplace to add to the coziness. The efficient kitchen has direct access to either the formal dining room or the breakfast room. Upstairs, the family's sleeping quarters give everyone their own private space. The master suite has a decorative ceiling and a walk-in closet. The master bath gives the home owner privacy and a little pampering. The three additional bedrooms share a full hall bath with a double vanity. A bonus room can be finished as the family's needs change. This plan is available with a basement, slab or crawl space foundation. Please specify when ordering. The photographed home has been modified to suit individual tastes.

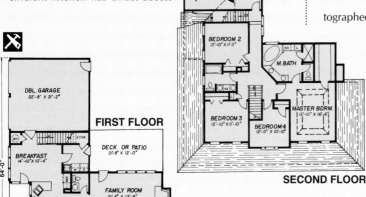

FIRST FLOOR

SECOND FLOOR

PLAN INFO

FIRST FL.	1,320 sq. ft.
SECOND FL.	1,268 sq. ft.
BASEMENT	1,320 sq. ft.
GARAGE	482 sq. ft.
BEDROOMS	Four
BATHROOMS	2(Full), 1(Half)
FOUNDATION	Basement, Slab, Crawl Space
TOTAL LIVING AREA	**2,588 sq. ft.**

An
EXCLUSIVE DESIGN
By Jannis Vann & Associates, Inc.

Refer To Price Code D on Page 253

DESIGN 93205

Everything You Need

The wrap-around covered porch and windows of this elevation combine to create a striking appearance, sure to attract attention from the curb. The entry offers a tremendous open view of the dining and Great room. The fireplace centers on the cathedral ceiling which soars to over sixteen feet high in the great room. The French doors to the dinette add a formal touch to the area. The kitchen features such amenities as a lazy Susan, a large food preparation island and an ample pantry. Double doors access the master bedroom which is further enhanced by a decorative boxed ceiling. The master bath has a large whirlpool, a separate shower, a makeup vanity and a walk-in closet. A walk-in closet is also featured in the third bedroom. The photographed home has been modified to suit individual tastes.

SECOND FLOOR

Br. 2
12⁰ x 12⁰

Br. 4
12⁰ x 11⁰

Br. 3
12⁰ x 11⁰

8'-8" CEILING

OPEN TO BELOW

PLANT SHELF

PLAN INFO

FIRST FL.	1,570 sq. ft.
SECOND FL.	707 sq. ft.
BASEMENT	1,570 sq. ft.
GARAGE	504 sq. ft.
BEDROOMS	Four
BATHROOMS	2(Full), 1(Half)
FOUNDATION	Basement
TOTAL LIVING AREA	**2,277 sq. ft.**

Refer To Price Code D on Page 253

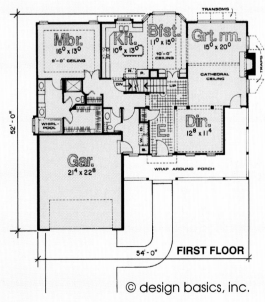

FIRST FLOOR

Mbr.
16⁰ x 13⁰
9'-0" CEILING

Kit.
10⁶ x 13⁰

Bfst.
11⁰ x 15⁰
10'-0" CEILING

Grt. rm.
15⁰ x 20⁰
CATHEDRAL CEILING

TRANSOMS

Din.
12⁸ x 11⁴

WHIRL POOL

Gar.
21⁴ x 22⁸

WRAP AROUND PORCH

52'-0"

54'-0"

© design basics, inc.

DESIGN 99431

Back to the Basics

Stacked windows fill the wall in the front bedroom of this one-level home, creating an attractive facade and a sunny atmosphere inside. Around the corner, two more bedrooms and two full baths complete the bedroom wing, set apart for bedtime quiet. Notice the elegant vaulted-ceiling in the master bedroom, the master tub and shower illuminated by a skylight, and the double vanities in both baths. Active areas enjoy a spacious feeling. Look at the high, sloping ceilings in the fireplaced living room, the sliders that unite the breakfast room and kitchen with an adjoining deck, and the vaulted-ceilings in the formal dining room off the foyer.

MAIN FLOOR

An
EXCLUSIVE DESIGN
By Karl Kreeger

PLAN INFO

MAIN FL.	1,737 sq. ft.
BASEMENT	1,727 sq. ft.
GARAGE	484 sq. ft.
BEDROOMS	Three
BATHROOMS	2(Full)
FOUNDATION	Basement, Slab, Crawl Space
TOTAL LIVING AREA	**1,737 sq. ft.**

Refer To Price Code B on Page 253

DESIGN 20100

Elegant Facade

This warm and inviting home features a see-through fireplace between the living room and family room. The gourmet kitchen gives the cook in your family the added work space of an island, plus all the amenities you've come to expect. Efficiently designed, the kitchen easily serves both the formal dining room and the nook. Upstairs, four bedrooms accommodate your sleeping hours. The master bedroom adds interest with a vaulted ceiling. The master bath has a large double vanity, linen closet, corner tub, separate shower, compartmented toilet, and huge walk-in closet. The three additional bedrooms, one with a walk-in closet, share the full hall bath.

SECOND FLOOR

OPTIONAL RETREAT

PLAN INFO

FIRST FL.	1,241 sq. ft.
SECOND FL.	1,170 sq. ft.
GARAGE	500 sq. ft.
BEDROOMS	Four
BATHROOMS	2(Full), 1(Half)
FOUNDATION	Basement, Slab, Crawl Space
TOTAL LIVING AREA	**2,411 sq. ft.**

Refer To Price Code D on Page 253

Refer To Price Code D on Page 253

An EXCLUSIVE DESIGN
By Energetic Enterprises

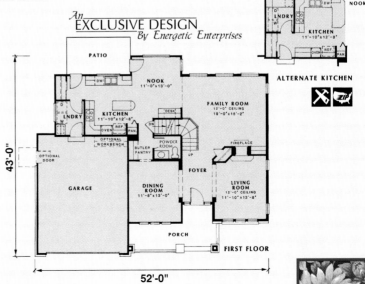

ALTERNATE KITCHEN

FIRST FLOOR

DESIGN 24262

Single Level Ease

The fireplace and sloped-ceiling in the family room offer something a bit out of the ordinary in this small home. The master bedroom is complete with a full bath and a dressing area. Bedrooms two and three share a full bath across the hall, and a half-bath is conveniently located adjacent to the kitchen. A bump-out bay window is shown in the spacious breakfast room, and a bay window with a window seat has been designed in the master bedroom. The screened porch off of the breakfast room is an inviting feature for meals outside.

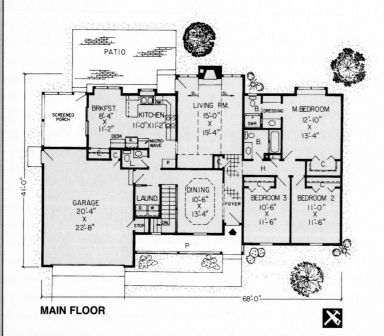

MAIN FLOOR

An
EXCLUSIVE DESIGN
By Karl Kreeger

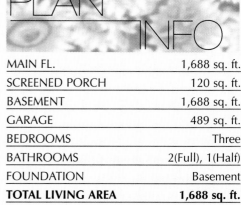

PLAN INFO

MAIN FL.	1,688 sq. ft.
SCREENED PORCH	120 sq. ft.
BASEMENT	1,688 sq. ft.
GARAGE	489 sq. ft.
BEDROOMS	Three
BATHROOMS	2(Full), 1(Half)
FOUNDATION	Basement
TOTAL LIVING AREA	**1,688 sq. ft.**

Refer To Price Code B on Page 253

DESIGN 10548

Sit Back & Relax

This delightful home's wrap-around covered porch recalls the warmth and charm of days past - lounging in the porch swing, savoring life. Inside, a spacious foyer welcomes guests and provides easy access to the formal dining room, secluded den/guest room (which might serve as your home office), and the large living room. Ceilings downstairs are all 9' high, with decorative vaults in the living and dining rooms. The kitchen, with its island/breakfast bar, is large enough for two people to work in comfortably. The adjacent laundry room also serves as a mud room for boots and clothes, and leads directly to the garage, which features an ample storage/shop area at the rear. Upstairs, three bedrooms, each with cathedral ceilings, share a cheery, sunlit sitting area that also features a cathedral ceiling. For privacy, the master bedroom is separated from the other bedrooms, and boasts a palatial bathroom, complete with a whirlpool tub. If room to relax is what you're after, this home is loaded with irresistible features.

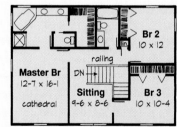

SECOND FLOOR

Br 2
10 x 12

Master Br
12-7 x 16-1
cathedral

railing

DN

Sitting
9-6 x 8-6

Br 3
10 x 10-4

PLAN INFO

FIRST FL.	1,034 sq. ft.
SECOND FL.	944 sq. ft.
BASEMENT	944 sq. ft.
GARAGE & STORAGE	675 sq. ft.
BEDROOMS	Three
BATHROOMS	2(Full), 1(Half)
FOUNDATION	Basement, Slab, Crawl Space
TOTAL LIVING AREA	**1,978 sq. ft.**

Refer To Price Code C on Page 253

FIRST FLOOR

Living
21-2 x 12-4
decor clg.

Kitchen
14-11 x 12-4

Storage/Shop
16-2 x 12-7

Den/Guest
10 x 10

Dining
10 x 12-3
decor clg.

Garage
23-2 x 19-3

39'-6"

67'-6"

crawl access

Dining

furn. w/h

DESIGN 24400

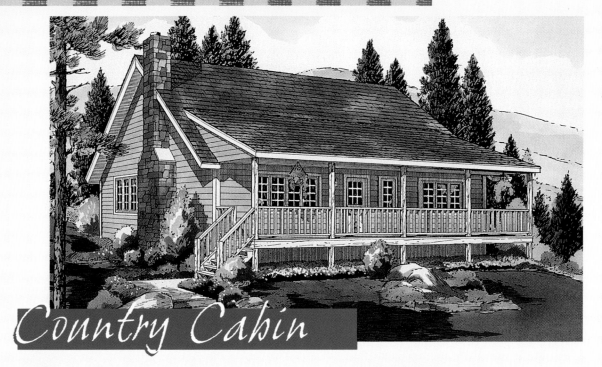

Country Cabin

Although rustic in appearance, the interior of this cabin is quiet, modern and comfortable. Small in overall size, it still contains three bedrooms and two baths in addition to a large, two-story living room with exposed beams. As a hunting/fishing lodge or mountain retreat, this compares well.

Open to Living Room Below

Flat Clg @ 7'-6"

Master Br
12-0 x 13-4

DN

Upper Floor

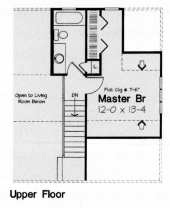

FURN   HH

Crawl Space Access

Crawl Space / Slab Plan

38'-0"

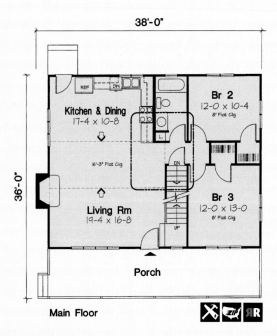

REF   DW

Kitchen & Dining
17-4 x 10-8

16'-3" Flat Clg

Br 2
12-0 x 10-4
8' Flat Clg

36'-0"

DN

Br 3
12-0 x 13-0
8' Flat Clg

Living Rm
19-4 x 16-8

UP

Porch

Main Floor

PLAN INFO

FIRST FL.	1,013 sq. ft.
SECOND FL.	315 sq. ft.
BASEMENT	1,013 sq. ft.
BEDROOMS	Three
BATHROOMS	2(Full)
FOUNDATION	Basement, Slab, Crawl Space
TOTAL LIVING AREA	**1,328 sq. ft.**

Refer To Price Code A on Page 253

Refer To Price Code A on Page 253

DESIGN 34600

Peaks Add Interest

From the welcoming porch to the balcony overlooking the skylit living room, this three bedroom beauty is loaded with sunny appeal. An elegant, bayed dining room adjoins the centrally located island kitchen, which features easy access to a screened porch. A short hall leads past the laundry and handy powder room to a huge, fireplaced living room that opens to a rear deck. Walk under the balcony to the first floor master suite with its walk-in closet and luxury bath. You'll love the quiet, yet convenient location of this special retreat. The second floor balcony overlooking the living room and two-story foyer links two more bedrooms, each with a huge closet, and a large divided bath.

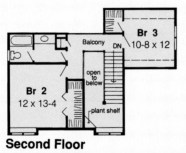

Second Floor

Br 3
10-8 x 12

Balcony   DN

open to below

Br 2
12 x 13-4

plant shelf

PLAN INFO

FIRST FL.	1,293 sq. ft.
SECOND FL.	526 sq. ft.
BASEMENT	1,286 sq. ft.
GARAGE	484 sq. ft.
BEDROOMS	Three
BATHROOMS	2(Full), 1(Half)
FOUNDATION	Basement
TOTAL LIVING AREA	**1,819 sq. ft.**

Refer To Price Code C on Page 253

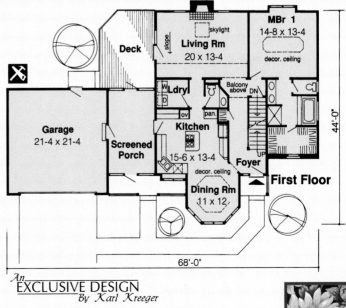

Deck

slope

skylight

Living Rm
20 x 13-4

MBr 1
14-8 x 13-4

decor. ceiling

W Ldry
D

ov

Balcony above   DN

Garage
21-4 x 21-4

Kitchen
15-6 x 13-4

pan.

Screened Porch

decor. ceiling

UP

Foyer

Dining Rm
11 x 12

First Floor

44'-0"

68'-0"

An
EXCLUSIVE DESIGN
By Karl Kreeger

DESIGN 20158

—twenty five—

Lots of Charm

alk past the charming front porch, in through the foyer and you'll be struck by the exciting, spacious living room. Complete with high sloping ceilings and a beautiful fireplace flanked by large windows. The large master bedroom shows off a full wall of closet space, its own private bath, and an extraordinary decorative ceiling. Just down the hall are two more bedrooms and another full bath. Take advantage of the accessibility off the foyer and turn one of these rooms into a private den or office space. The dining room provides a feast for your eyes with its decorative ceiling details, and a full slider out to the deck. Along with great counter space, the kitchen includes a double sink and an attractive bump-out window. The adjacent laundry room, optional expanded pantry, and a two-car garage make this Ranch a charmer.

Crawl Space Access

Slab/Crawl Space Option

An
EXCLUSIVE DESIGN
By Karl Kreeger

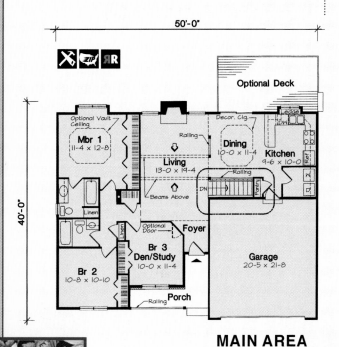

50'-0"

40'-0"

Optional Vault Ceiling

Mbr 1
11-4 x 12-8

Optional Deck

Decor. Clg.

Dining
10-0 x 11-4

Kitchen
9-6 x 10-0

Railing

Living
13-0 x 19-4

Railing

Beams Above

DN

Linen

Pantry

Optional Door

Foyer

Br 3
Den/Study
10-0 x 11-4

Br 2
10-8 x 10-10

Garage
20-5 x 21-8

Porch

Railing

MAIN AREA

PLAN INFO

MAIN FL.	1,307 sq. ft.
BASEMENT	1,298 sq. ft.
GARAGE	462 sq. ft.
BEDROOMS	Three
BATHROOMS	2(Full)
FOUNDATION	Basement, Slab, Crawl Space
TOTAL LIVING AREA	**1,307 sq. ft.**

Refer To Price Code A on Page 253

DESIGN 20161

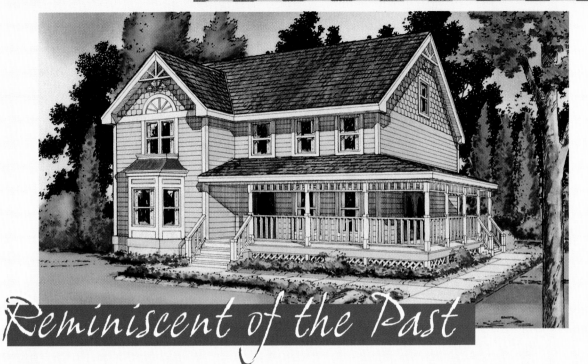

Reminiscent of the Past

R eminiscent of a simpler time, yet up-to-date on the conveniences we have come to depend on. The front porch is a warm welcome at the end of the day. Come inside to the large living room, great for entertaining. Or relax in your family room which opens to your deck. Upstairs the master bedroom has its own bath and ample closet space. The secondary bedrooms, of which there are three, share the hall bath and each have lots of closet space. Family living at its best!

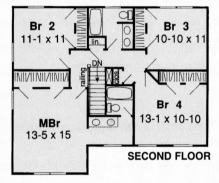

Br 2 11-1 x 11

Br 3 10-10 x 11

lin.

DN

railing

MBr 13-5 x 15

Br 4 13-1 x 10-10

SECOND FLOOR

An
EXCLUSIVE DESIGN
By Marshall Associates

PLAN INFO

FIRST FL.	987 sq. ft.
SECOND FL.	970 sq. ft.
BASEMENT	985 sq. ft.
BEDROOMS	Four
BATHROOMS	2(Full), 1(Half)
FOUNDATION	Basement
TOTAL LIVING AREA	**1,957 sq. ft.**

Refer To Price Code C on Page 253

Refer To Price Code C on Page 253

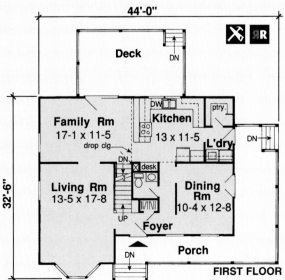

44'-0"

Deck

DN

Family Rm 17-1 x 11-5

DW

Kitchen 13 x 11-5

ptry.

drop clg.

DN

L'dry

DN

32'-6"

DN

desk

Living Rm 13-5 x 17-8

Dining Rm 10-4 x 12-8

UP

Foyer

DN

Porch

FIRST FLOOR

DESIGN 24301

Maximum Appeal

A large front porch is always an old-fashioned welcome to any home. This Cape provides such a welcome. Once inside the home, the vaulted ceiling and grand fireplace of the living room add to the character of this house. The efficient kitchen has a double sink and a peninsula counter that may double as an eating bar. A laundry center is conveniently located in the full bath. Two of the three bedrooms are located on the first floor. The second floor provides the privacy the master suite deserves. Sloping ceilings, a walk-in closet and a private master bath give the owner of this home a private retreat on the second floor.

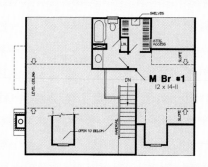

Second Floor

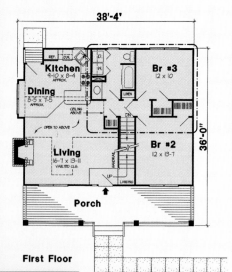

First Floor

Crawl Space Option

PLAN INFO

FIRST FL.	1,007 sq. ft.
SECOND FL.	408 sq. ft.
BASEMENT	1,007 sq. ft.
BEDROOMS	Three
BATHROOMS	2(Full)
FOUNDATION	Basement, Slab, Crawl Space
TOTAL LIVING AREA	**1,415 sq. ft.**

Refer To Price Code A on Page 253

DESIGN 34601

Family Living

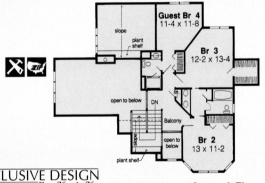

Second Floor

Guest Br 4
11-4 x 11-8

slope

plant shelf

Br 3
12-2 x 13-4

open to below

DN

Balcony

open to below

Br 2
13 x 11-2

plant shelf

rom the sprawling front porch to the two-way fireplace that warms the hearth room and living room, this house says "Welcome" to all who enter. Even your houseplants will love the cozy, sunny atmosphere of this country classic. A central hallway links the formal, bayed dining room with the spacious kitchen at the rear of the house. Relax over an informal meal in the adjoining hearth room, or out on the deck when the weather's warm. It's accessible from both the hearth room and soaring, wide-open living room. Relish the privacy of your first floor master suite, which features a bath with every amenity, and a huge, walk-in closet. Step upstairs for a great view of active areas, where you'll find three more bedrooms, each adjoining a full bath.

An
EXCLUSIVE DESIGN
By Karl Kreeger

PLAN INFO

FIRST FL.	1,737 sq. ft.
SECOND FL.	826 sq. ft.
BASEMENT	1,728 sq. ft.
BEDROOMS	Four
BATHROOMS	3(Full), 1(Half)
FOUNDATION	Basement
TOTAL LIVING AREA	**2,563 sq. ft.**

Refer To Price Code D on Page 253

61'-0"

52'-0"

Deck

Hearth Rm
13-4 x 14-8

slope

Kit
11-4 x 12

W D

Ldry

FZR

Garage
21-8 x 21-4

Living Rm
13-8 x 22
17'-0" ceiling height

UP DN

Balcony above

Foyer

Dining Rm
13 x 13-6

plant shelf

MBr 1
14-4 x 15-4
ceiling vaulted

First Floor

DESIGN 20144

Quaint Charmer

The exterior of this Ranch home is all wood with interesting lines. More than an ordinary ranch home, it has an expansive feeling to drive up to. The large living area has a stone fireplace and decorative beams. The kitchen and dining room lead to an outside deck. The laundry room has a large pantry, and is off the eating area. The master bedroom has a wonderful bathroom with a huge walk-in closet. In the front of the house, there are two additional bedrooms with a bathroom. This house offers one floor living and has nice big rooms.

56'-0"

Deck

Kitchen
12 x 11-4

Dining Rm
9 x 11-4

pantry

Ldry

MBr 1
14-2 x 14-4

Living Rm
21-6 x 19-4

decor. beams

Br 3
12 x 12-6

Br 2
12 x 12-6

lin.

32'-0"

slope

MAIN AREA

An
EXCLUSIVE DESIGN
By Karl Kreeger

PLAN INFO

MAIN FL.	1,792 sq. ft.
BASEMENT	818 sq. ft.
GARAGE	857 sq. ft.
BEDROOMS	Three
BATHROOMS	2(Full)
FOUNDATION	Basement
TOTAL LIVING AREA	**1,792 sq. ft.**

Refer To Price Code B on Page 253

DESIGN 20198

Country Bliss

The large, welcoming, wrap-around porch of this home may add a touch of an old-fashion country feel, but don't be fooled. This is a totally modern home. Through the decorative front door is a large entrance foyer with an attractive stair case to the second floor. The study/guest room is the left and has convenient access to a full hall bath. To the right of the foyer is the elegant dining room. A decorative ceiling treatment enhances the room. The family room includes a massive fireplace with built-in bookshelves. Built-in features continue in the breakfast room and kitchen. The breakfast room has a convenient built-in planning desk and the kitchen includes a built-in

pantry. The peninsula counter/eating bar separates the two rooms. A corner double sink and ample counter and storage space add to the efficiency of the kitchen. The shop area enters the home through the utility room keeping the rest of the home clean. The master bedroom is crowned by a cathedral ceiling and includes a private, lavish bath with a walk-in closet. Two additional bedrooms share a compartmented, double vanitied bath. There is an option to have a fourth bedroom.

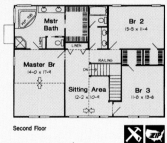

Optional Second Floor

Second Floor

PLAN INFO

FIRST FL.	1,378 sq. ft.
SECOND FL.	1,269 sq. ft.
BASEMENT	1,378 sq. ft.
GARAGE	717 sq. ft.
BEDROOMS	Three
BATHROOMS	2(Full), 1(Three-Quarter)
FOUNDATION	Basement, Slab, Crawl Space
TOTAL LIVING AREA	**2,647 sq. ft.**

Refer To Price Code E on Page 253

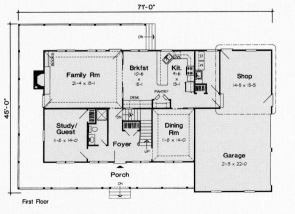

First Floor

Crawl Space/Slab Option

DESIGN 24403

Farmhouse Flavor

The charm of an old-fashioned farmhouse combines with sizzling contemporary excitement in this three-bedroom home. Classic touches abound, from the clapboard exterior with its inviting, wraparound porch to the wood stove that warms the entire house. Inside, the two-story foyer, crowned by a plant ledge high overhead, affords a view of the soaring, skylit living room and rear deck beyond sliding glass doors. To the right, there's a formal dining room with bay window, just steps away from the kitchen. The well-appointed master suite completes the first floor. Upstairs, you'll find a full bath and two more bedrooms, each with a walk-in closet and cozy gable sitting nook.

FIRST FLOOR

Optional Deck

Living Rm 13 x 19-6

Ldry

MBr 1 13-6 x 14

wood stove

Kitchen 11 x 12

DN

Dining Rm 12-10 x 13-6

Foyer

39'-0"

47'-0"

Slab/Crawl Space Option

SECOND FLOOR

slope / skylight / open to below

Balcony

Br 2 10-4 x 14

DN

Br 3 11 x 14

plant ledge

slope

An
EXCLUSIVE DESIGN
By Karl Kreeger

PLAN INFO

FIRST FL.	1,269 sq. ft.
SECOND FL.	638 sq. ft.
BASEMENT	1,269 sq. ft.
BEDROOMS	Three
BATHROOMS	2(Full), 1(Half)
FOUNDATION	Basement, Slab, Crawl Space
TOTAL LIVING AREA	**1,907 sq. ft.**

Refer To Price Code C on Page 253

DESIGN 10785

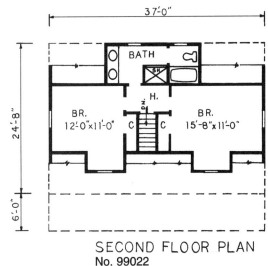

SECOND FLOOR PLAN
No. 99022

37'-0"

24'-8"

6'-0"

BATH

BR.
12'-0"x11'-0"

BR.
15'-8"x11'-0"

H.
DN. C C

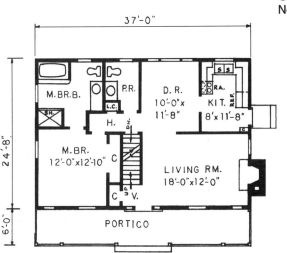

FIRST FLOOR PLAN

37'-0"

24'-8"

6'-0"

M.BR.B. P.R.

D.R.
10'-0"x
11'-8"

KIT.
8'x11'-8"

R.A.

REF.

M.BR.
12'-0"x12'-10"

LIVING RM.
18'-0"x12'-0"

H.
DN.
C
C V.

L.C.

SH

PORTICO

TOTAL LIVING AREA:
1,494 SQ. FT.

Three Bedroom Traditional Country Cape

■ This plan features:

— Three bedrooms

— Two full and one half baths

■ Entry area with a coat closet

■ Ample sized Living Room with a fireplace

■ Dining Room with a view of the rear yard and located conveniently close to the Kitchen and Living Room

■ U-shaped Kitchen with a double sink, ample cabinet and counter space and a side door to the outside

■ First floor Master Suite with a private Master Bath

■ Two additional, second floor bedrooms that share a full, double vanity bath with a separate shower

FIRST FLOOR — 913 SQ. FT.
SECOND FLOOR — 581 SQ. FT.

Refer to **Pricing Schedule B** on the order form for pricing information

Cozy Front Porch

■ This plan features:

— Three bedrooms

— Two full and one half bath

■ Living Room enhanced by a large fireplace

■ Formal Dining Room that is open to the Living Room, giving a more spacious feel to the rooms

■ Efficient Kitchen with ample counter and cabinet space, double sinks and pass thru window to living area

■ Sunny Breakfast Area with vaulted ceiling and a door to the sun deck

■ First floor Master Suite with separate tub and shower stall, and walk-in closet

■ First floor powder room with a hide-away laundry center

■ Two additional bedrooms that share a full hall bath

An
EXCLUSIVE DESIGN
By Jannis Vann & Associates, Inc.

FIRST FLOOR — 1,045 SQ. FT.
SECOND FLOOR — 690 SQ. FT.
BASEMENT — 465 SQ. FT.
GARAGE — 580 SQ. FT.

TOTAL LIVING AREA:
1,735 SQ. FT.

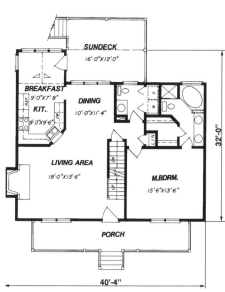

FIRST FLOOR
No. 93269

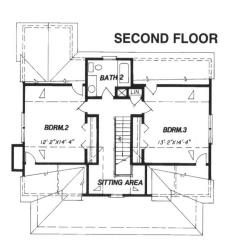

SECOND FLOOR

Ranch with Handicapped Access

■ This plan features:

— Three bedrooms

— Two full baths

■ Ramps into the front Entry from the Porch; the Utility area and the Kitchen from the Garage; and the Family Room from the Deck

■ An open area topped by a sloped ceiling for the Family Room, the Dining Room, the Kitchen and the Breakfast alcove

■ An efficient Kitchen, with a built-in pantry, easily serves both the Breakfast nook and the Dining Room

■ A Master Bedroom suite accented by a sloping ceiling above a wall of windows and access to the Deck

■ Two front bedrooms with sloped ceilings sharing a full hall bath

MAIN FLOOR — 1,734 SQ. FT.
PORCH — 118 SQ. FT.
DECK — 354 SQ. FT.
GARAGE — 606 SQ. FT.

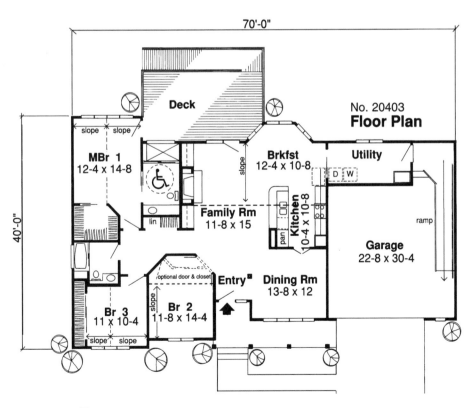

No. 20403
Floor Plan

TOTAL LIVING AREA:
1,734 SQ. FT.

Refer to **Pricing Schedule D** on the order form for pricing information

© 1997 Donald A. Gardner Architects, Inc.

B. NATHAN

Comfortable Design Encourages Relaxation

■ This plan features:

— Four bedrooms

— Three full baths

■ A wide front porch providing a warm welcome

■ Center dormer lighting foyer as columns punctuate the entry to the Dining Room and Great Room

■ Spacious Kitchen with angled countertop is open to the Breakfast Bay

■ Tray ceilings adding elegance to the Dining Room and Master Bedroom

■ Master Suite located on the opposite end of the home; and an optional arrangement for the physically challenged

■ Two bedrooms share a third full bath with a linen closet

■ Skylight Bonus room is located over the garage

MAIN AREA — 2,349 SQ. FT.
GARAGE — 615 SQ. FT.

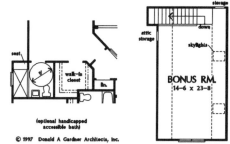

(optional handicapped accessible bath)

© 1997 Donald A Gardner Architects, Inc.

BONUS RM.
14–6 x 23–8

TOTAL LIVING AREA:
2,349 SQ. FT.

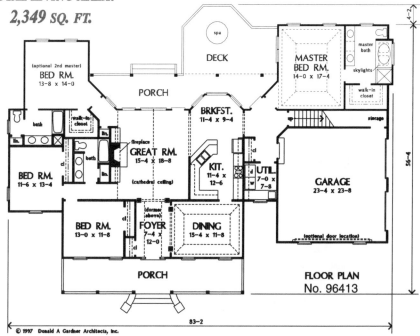

FLOOR PLAN
No. 96413

© 1997 Donald A Gardner Architects, Inc.

To order your Blueprints, call 1-800-235-5700

Refer to **Pricing Schedule A** on the order form for pricing information

ALTERNATE FLOOR PLAN for Crawl Space

D   W

Optional Deck

| Kit 9-8 x 10-1 | Brkfst 8-4 x 10-1 | Br 3 9-1 x 10-1 | Br 2 11-6 x 9-3 |

DN

26'-0"

Living Rm 17-0 x 11-6

fireplace

MBr 1 11-6 x 10-11

lin

Deck

MAIN AREA No. 34328

42'-0"

Compact Ranch Loaded with Living Space

■ This plan features:

— Three bedrooms

— One full bath

■ A central entrance, opening to the Living Room with ample windows

■ A Kitchen, featuring a Breakfast area with sliding doors to the backyard and an optional deck

MAIN AREA — 1,092 SQ. FT.
BASEMENT — 1,092 SQ. FT.

TOTAL LIVING AREA:
1,092 SQ. FT.

To order your Blueprints, call 1-800-235-5700

Arched Windows Add Natural Light

◼ This plan features:

— Four bedrooms

— Three full baths

◼ Dining Room highlighted by alcove of windows

◼ Spacious Living Room enhanced by a hearth fireplace, built-in shelves and glass access to Covered Porch

◼ Efficient, U-shaped Kitchen with a peninsula serving bar, nearby laundry and bright Breakfast area with access to Porch

◼ Pampering bedroom one offers a large walk-in closet and double vanity bath

◼ Three additional bedrooms with walk-in closets, have access to full baths

◼ No materials list available

◼ Please specify a crawl space or slab foundation when ordering

FIRST FLOOR — 1,505 SQ. FT.
SECOND FLOOR — 555 SQ. FT.

TOTAL LIVING AREA:
2,060 SQ. FT.

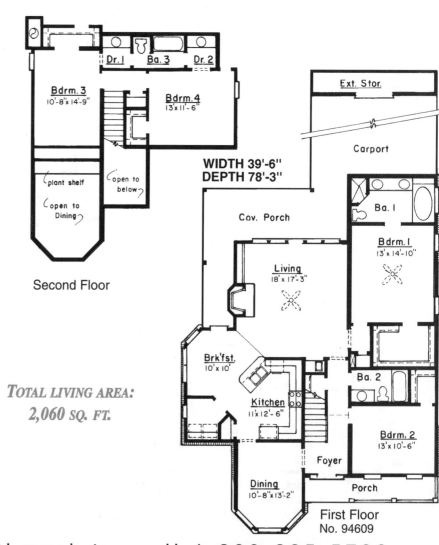

WIDTH 39'-6"
DEPTH 78'-3"

Second Floor

First Floor
No. 94609

Entertaining at its Best

■ This plan features:

— Three bedrooms

— Two full and one half baths

■ A Master Bedroom privately set with a sitting area, full bath and walk-in closet

■ An island Kitchen centered between the Dining Room and Breakfast area

■ A sunken Living Room with vaulted ceilings and a two-way fireplace

■ A covered porch and enormous deck

FIRST FLOOR — 1,818 SQ. FT.
SECOND FLOOR — 528 SQ. FT.
BASEMENT — 1,818 SQ. FT.

TOTAL LIVING AREA:
2,346 SQ. FT.

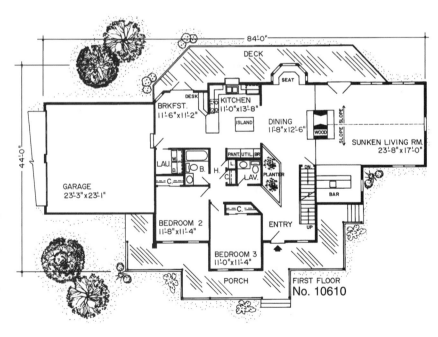

Refer to **Pricing Schedule E** on the order form for pricing information

Victorian Details

■ This plan features:

— Four bedrooms

— Two full and one half baths

■ A large country Kitchen in full view of a breakfast area

■ A fireplace shared by the cozy Living Room and the Family Room containing a bar and access to the patio

■ Octagonal recessed ceilings in the formal Dining Room

■ Walk-in closets enhancing all the bedrooms

FIRST FLOOR — 1,450 SQ. FT.
SECOND FLOOR — 1,341 SQ. FT.
BASEMENT — 1,450 SQ. FT.
GARAGE — 629 SQ. FT.
COVERED PORCH — 144 SQ. FT.
WOOD STORAGE — 48 SQ. FT.

TOTAL LIVING AREA:
2,791 SQ. FT.

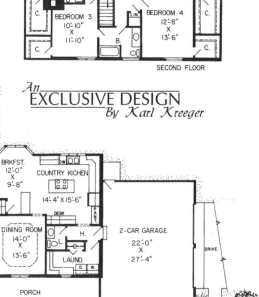

An *EXCLUSIVE DESIGN By Karl Kreeger*

First Floor
No. 10593

Refer to **Pricing Schedule C** on the order form for pricing information

TOTAL LIVING AREA:
1,367 SQ. FT.

An
EXCLUSIVE DESIGN
By Jannis Vann & Associates, Inc.

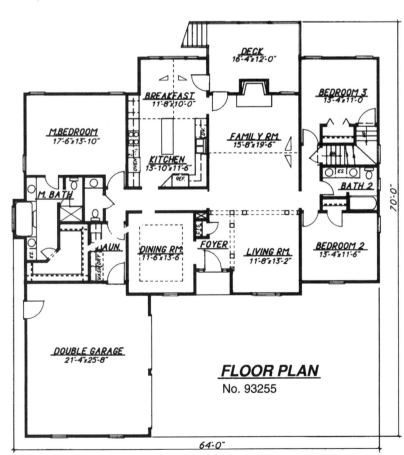

FLOOR PLAN
No. 93255

One Story Country Home

▪ This plan features:

— Three bedrooms

— Two full and one half baths

▪ A modern, convenient floor plan

▪ Formal areas located at the front of the home

▪ A decorative ceiling in the Dining Room

▪ Columns accent the Living Room

▪ A large Family Room with a cozy fireplace and direct access to the deck

▪ An efficient Kitchen located between the formal Dining Room and the informal Breakfast Room

▪ A private Master Suite with a Master Bath and walk-in closet

▪ Two additional bedrooms share a full hall bath

MAIN AREA — 2,192 SQ. FT.
BASEMENT — 2,192 SQ. FT.
GARAGE — 564 SQ. FT.

Refer to **Pricing Schedule C** on the order form for pricing information

© 1992 Donald A. Gardner Architects, Inc.

B·NATHAN·

Rustic Three Bedroom

■ This plan features:

— Three bedrooms

— Two full baths

■ Spacious covered porches on the front and rear of the home

■ The open Great Room provides a spacious feeling of a much large home

■ Cooktop island Kitchen includes L-shaped counter for ample work space

■ Master suite has a generous walk-in closet and pampering master bath

■ Two second floor bedrooms, one over looking the Great Room for added drama

FIRST FLOOR — 1,039 SQ. FT.
SECOND FLOOR — 583 SQ. FT.

TOTAL LIVING AREA: 1,622 SQ. FT.

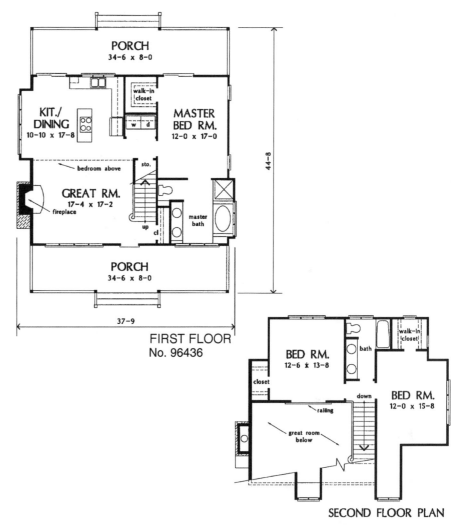

PORCH
34-6 x 8-0

walk-in closet

KIT./ DINING
10-10 x 17-8

MASTER BED RM.
12-0 x 17-0

w d

bedroom above

sto.

GREAT RM.
17-4 x 17-2
fireplace

up

cl

master bath

PORCH
34-6 x 8-0

44-8

37-9

FIRST FLOOR
No. 96436

BED RM.
12-6 x 13-8

walk-in closet

bath

closet

down

railing

great room below

BED RM.
12-0 x 15-8

SECOND FLOOR PLAN

To order your Blueprints, call 1-800-235-5700

An
EXCLUSIVE DESIGN
By Jannis Vann & Associates, Inc.

Country Victorian

■ This plan features:

— Four bedrooms

— Two full and one half baths

■ Victorian accent on a country porch with an octagonal sitting area

■ A central Foyer area highlighted by a decorative staircase and a convenient coat closet

■ An elegant bay window enhancing the formal Dining Room

■ A formal Living Room viewing the porch and front yard

■ A large country Kitchen with a central work island and an open layout into the cheery Breakfast Bay

■ An expansive Family Room equipped with a cozy fireplace and two sets of French doors

■ Master Suite with a sitting alcove, walk-in closet, bath and a private balcony

Second Floor

FIRST FLOOR — 1,155 SQ. FT.
SECOND FLOOR — 1,209 SQ. FT.
BASEMENT — 549 SQ. FT.
GARAGE — 576 SQ. FT.

TOTAL LIVING AREA:
2,364 SQ. FT.

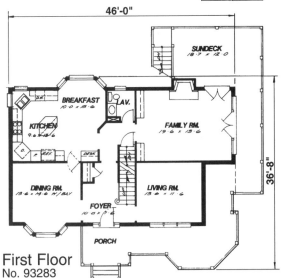

First Floor
No. 93283

© 1994 Donald A. Gardner Architects, Inc.

Refer to **Pricing Schedule E** on the order form for pricing information

The Great Outdoors

- This plan features:
- — Four bedrooms
- — Two full and one half baths
- Bay windows and a long, skylit, screened Porch — a haven for outdoor enthusiasts
- Open foyer takes advantage of light from central dormer with palladian window
- Vaulted ceiling in the Great Room adds vertical drama
- Contemporary Kitchen is open to the Great Room creating a spacious feeling
- Master Bedroom Suite is privately tucked away with a large luxurious bath complete with a bay window, corner shower, and a garden tub

FIRST FLOOR — 1,907 SQ. FT.
SECOND FLOOR — 656 SQ. FT.
BONUS ROOM — 467 SQ. FT.
GARAGE & STORAGE — 580 SQ. FT.

TOTAL LIVING AREA:
2,563 SQ. FT.

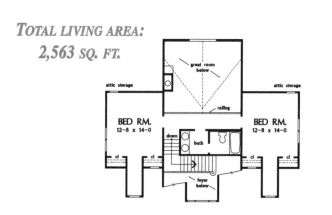

SECOND FLOOR PLAN

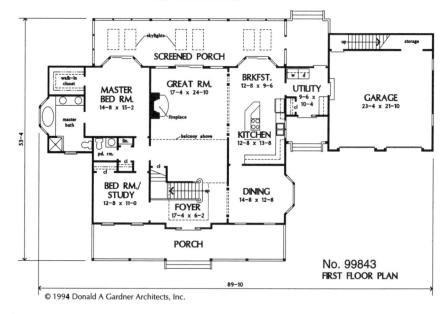

No. 99843
FIRST FLOOR PLAN

© 1994 Donald A Gardner Architects, Inc.

To order your Blueprints, call 1-800-235-5700

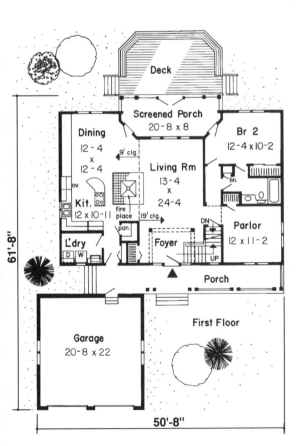

First Floor

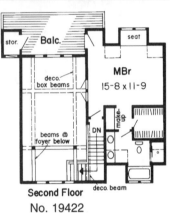

Second Floor
No. 19422

FIRST FLOOR — 1,290 SQ. FT.
SECOND FLOOR — 405 SQ. FT.
SCREENED PORCH — 152 SQ. FT.
GARAGE — 513 SQ. FT.

TOTAL LIVING AREA:
1,695 SQ. FT.

Master Retreat Crowns Spacious Home

■ This plan features:

— Two bedrooms

— Two full baths

■ An open Foyer leading up an landing staircase with windows above and into a two-story Living Room

■ A unique four-sided fireplace separates the Living Room, Dining area and Kitchen

■ A well-equipped Kitchen featuring a cook island, a walk-in pantry and easy access to Dining area and Laundry room

■ A three season Screened Porch and Deck beyond adjoining Dining Room, Living Room, and second Bedroom

■ An private Master Suite on the second floor offering a cozy, dormer window seat, private balcony, and window tub in the spacious Bath

Refer to **Pricing Schedule C** on the order form for pricing information

Country Living in a Doll House

■ This plan features:

— Three bedrooms

— Two full and one half baths

■ An eat-in country Kitchen with an island counter and bay window

■ A spacious Great Room with a fireplace flowing easily into the Dining area

■ A first floor Master Suite including a walk-in closet and a private compartmentalized bath

■ Two additional bedrooms sharing a full bath with a double vanity

■ An optional basement or crawl space foundation — please specify when ordering

FIRST FLOOR — 1,277 SQ. FT.
SECOND FLOOR — 720 SQ. FT.

TOTAL LIVING AREA: 1,997 SQ. FT.

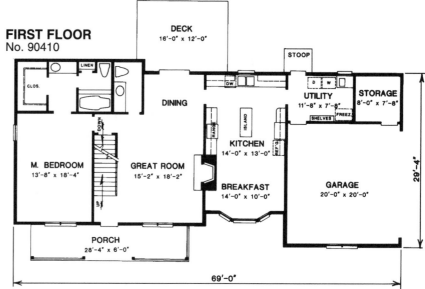

FIRST FLOOR
No. 90410

DECK
16'-0" x 12'-0"

STOOP

LINEN

CLOS.

DOWN

M. BEDROOM
13'-8" x 18'-4"

GREAT ROOM
15'-2" x 18'-2"

DINING

DW

UTILITY
11'-8" x 7'-8"

STORAGE
8'-0" x 7'-8"

SHELVES FREEZ.

RANGE

ISLAND

KITCHEN
14'-0" x 13'-0"

REF'G

BREAKFAST
14'-0" x 10'-0"

GARAGE
20'-0" x 20'-0"

29'-4"

PORCH
28'-4" x 6'-0"

69'-0"

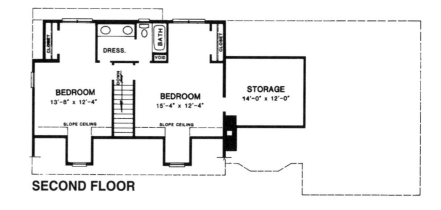

CLOSET

BATH

DRESS.

VOID

CLOSET

BEDROOM
13'-8" x 12'-4"

DOWN

BEDROOM
15'-4" x 12'-4"

STORAGE
14'-0" x 12'-0"

SLOPE CEILING

SLOPE CEILING

SECOND FLOOR

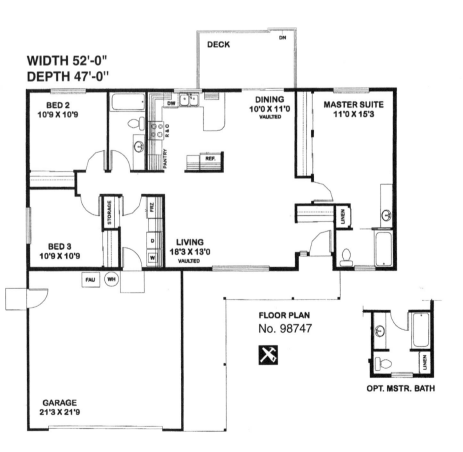

WIDTH 52'-0"
DEPTH 47'-0"

DECK — DN

BED 2
10'9 X 10'9

DW

PANTRY — R & O

REF.

DINING
10'0 X 11'0
VAULTED

MASTER SUITE
11'0 X 15'3

STORAGE

FRZ

LINEN

BED 3
10'9 X 10'9

D

W

LIVING
18'3 X 13'0
VAULTED

FAU   WH

FLOOR PLAN
No. 98747

GARAGE
21'3 X 21'9

LINEN

OPT. MSTR. BATH

L-Shaped Front Porch

■ This plan features:

— Three bedrooms

— Two full baths

■ Attractive wood siding and a
large L-shaped covered porch

■ Generous Living Room with a
vaulted ceiling

■ Large two car garage with access
through Utility Room

■ Roomy secondary bedrooms
share the full bath in the hall

■ Kitchen highlighted by a built-in
pantry and a garden window

■ Vaulted ceiling adds volume to
the Dining Room

■ Master Suite in isolated location
enhanced by abundant closet
space, separate vanity, and linen
storage

MAIN FLOOR — 1,280 SQ. FT.

TOTAL LIVING AREA:
1,280 SQ. FT.

Refer to **Pricing Schedule D** on the order form for pricing information

First Floor Master Suite

- This plan features:
- — Four bedrooms
- — Three full and one half baths
- Welcoming country porch adds to appeal and living space
- Spacious Living Room with fireplace and access to Covered Porch
- Efficient Kitchen with work island, built-in pantry, Utility room, Garage entry and Breakfast area
- Spacious Master Bedroom suite with a pampering, private bath
- Three second floor bedrooms with walk-in closets, share two full baths and a Game Room
- No materials list available
- Please specify a crawl space or slab foundation when ordering

FIRST FLOOR — 1,492 SQ. FT.
SECOND FLOOR — 865 SQ. FT.
BONUS — 303 SQ. FT.
GARAGE — 574 SQ. FT.

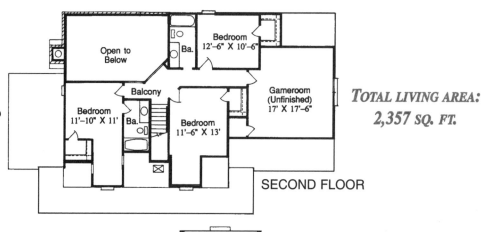

TOTAL LIVING AREA:
2,357 SQ. FT.

SECOND FLOOR

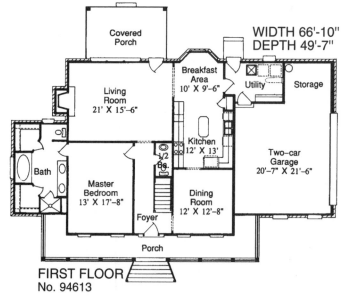

WIDTH 66'-10"
DEPTH 49'-7"

FIRST FLOOR
No. 94613

To order your Blueprints, call 1-800-235-5700

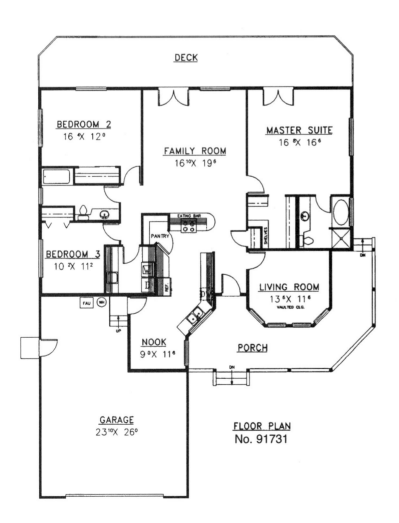

DECK

BEDROOM 2
16⁴X 12⁰

FAMILY ROOM
16¹⁰X 19⁶

MASTER SUITE
16⁶X 16⁶

EATING BAR

PANTRY

BEDROOM 3
10²X 11²

SHELVES

REF.

FAU   WH

LIVING ROOM
13⁶X 11⁶
VAULTED CLG.

UP

NOOK
9⁰X 11⁶

PORCH

DN

GARAGE
23¹⁰X 26⁰

FLOOR PLAN
No. 91731

Country Style & Charm

■ This plan features:

— Three bedrooms

— Two full baths

■ Brick accents, front facing gable, and railed wrap-around covered porch

■ A built-in range and oven in a dog-leg shaped Kitchen

■ A Nook with garage access for convenient unloading of groceries and other supplies

■ A bay window wrapping around the front of the formal Living Room

■ A Master Suite with French doors opening to the deck

MAIN AREA — 1,857 SQ. FT.
GARAGE — 681 SQ. FT.
WIDTH — 51'-6"
DEPTH — 65'-0"

TOTAL LIVING AREA:
1,857 SQ. FT.

Refer to **Pricing Schedule A** on the order form for pricing information

An EXCLUSIVE DESIGN
By Karl Kreeger

Simple Lines Enhanced by Elegant Window Treatment

■ This plan features:

— Two bedrooms (optional third)

— Two full baths

■ A huge, arched window that floods the front room with natural light

■ A homey, well-lit Office or Den

■ Compact, efficient use of space

■ An efficient Kitchen with easy access to the Dining Room

■ A fireplaced Living Room with a sloping ceiling and a window wall

■ A Master Bedroom sporting a private Master Bath with a roomy walk-in closet

MAIN AREA — 1,492 SQ. FT.
BASEMENT — 1,486 SQ. FT.
GARAGE — 462 SQ. FT.

TOTAL LIVING AREA:
1,492 SQ. FT.

56'-0"

Deck (Optional)

Optional Clg Reveal

Dining
10-10 x 11-4

Living Rm
14-6 x 20-10

Sloped Ceiling

W.P. Tub

Step

Master Br
13-8 x 13-6

Desk

DW

Kit.
10-10 x 10-0

P.

Ref

Railing

Pantry

DN

Den/
Br #3
10-6 x 12-0
Flat Clg
@ 10'

Br #2
13-8 x 11-6

48'-0"

MAIN AREA
No. 34150

Garage
20-5 x 21-8

W. D. HW Furn

Slab/Crawlspace Option

© 1993 Donald A. Gardner Architects, Inc.

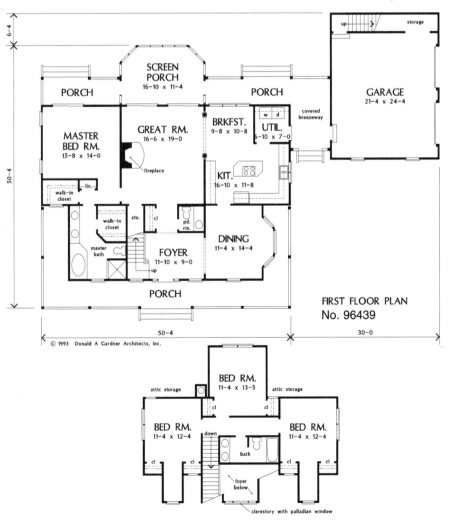

Farmhouse with Charm and Style

■ This plan features:

— Four bedrooms

— Two full and one half baths

■ Arched windows, dormers and expansive porches giving charm and style

■ Nine foot ceiling on the first level expanding space visually

■ Great Room opening to a spacious screened porch

■ Well planned Kitchen with an island cooktop and ample cabinet space

■ Deluxe Master Suite with a private bath and two walk-in closets

■ Bonus room adding extra storage or living space

FIRST FLOOR — 1,585 SQ. FT.
SECOND FLOOR — 723 SQ. FT.
BONUS ROOM — 419 SQ. FT.
GARAGE & STORAGE — 594 SQ. FT.

TOTAL LIVNG AREA:
2,308 SQ. FT.

FIRST FLOOR PLAN
No. 96439

© 1993 Donald A Gardner Architects, Inc.

SECOND FLOOR PLAN

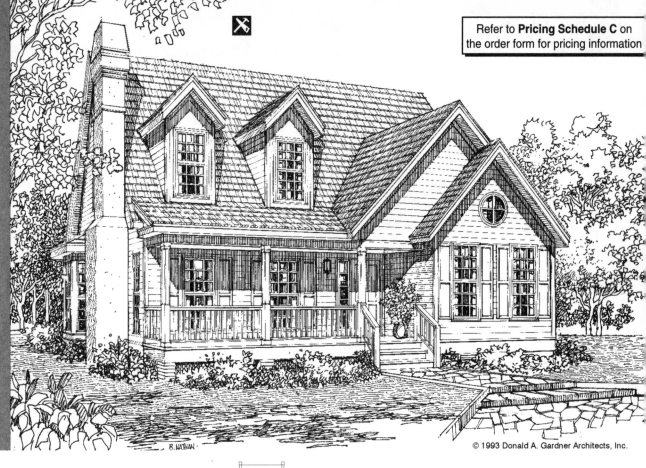

Refer to **Pricing Schedule C** on the order form for pricing information

© 1993 Donald A. Gardner Architects, Inc.

B. NATHAN

For Four or More

◾ This plan features:

— Three bedrooms

— Two full baths

◾ Covered porches, front and back, and an open interior capped by a cathedral ceiling

◾ Cathedral ceiling turns the Great Room, Kitchen/Dining and loft/study into impressive living spaces

◾ Kitchen equipped with a cooktop island and counter opening to both the Great Room and Dining Room

◾ A cathedral ceiling topps the front bedroom/study,

◾ A large bay cozies up the rear bedroom

◾ Luxurious Master Suite located upstairs for extra privacy

FIRST FLOOR — 1,146 SQ. FT.
SECOND FLOOR — 567 SQ. FT.

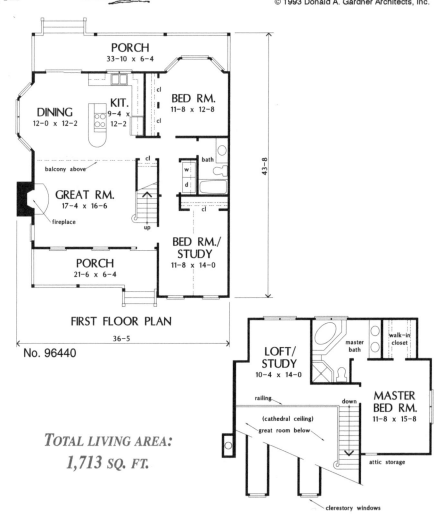

TOTAL LIVING AREA:
1,713 SQ. FT.

FIRST FLOOR PLAN

No. 96440

SECOND FLOOR PLAN

To order your Blueprints, call 1-800-235-5700

Refer to **Pricing Schedule E** on the order form for pricing information

An EXCLUSIVE DESIGN
By Westhome Planners, Ltd.

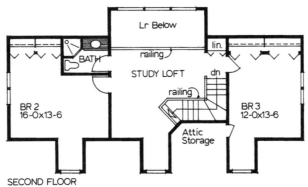

SECOND FLOOR

- Lr Below
- railing
- lin.
- BATH
- STUDY LOFT
- dn
- railing
- BR 2 16-0x13-6
- BR 3 12-0x13-6
- Attic Storage

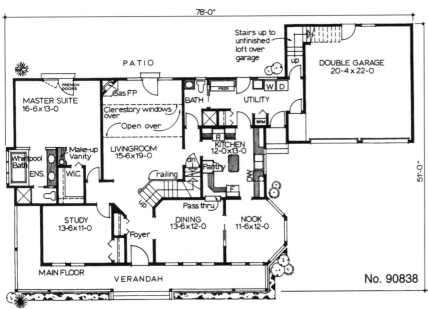

78'-0"

51'-0"

PATIO
Stairs up to unfinished loft over garage
up
DOUBLE GARAGE 20-4 x 22-0
FRENCH DOORS
MASTER SUITE 16-6x13-0
Gas FP
Clerestory windows over
Open over
BATH
FRZR
W D
UTILITY
Whirlpool Bath
ENS.
Make-up Vanity
W.I.C.
LIVINGROOM 15-6x19-0
railing
dn
KITCHEN 12-0x13-0
R
BRM
Pantry
DW
F
Pass thru
STUDY 13-6x11-0
Foyer
DINING 13-6x12-0
NOOK 11-6x12-0
MAIN FLOOR   VERANDAH

No. 90838

Room to Grow

◼ This plan features:

— Three bedrooms

— Three full baths

◼ A corner gas fireplace in the spacious Living Room

◼ A Master Suite including a private Bath with a whirlpool tub, separate shower and a double vanity

◼ An island Kitchen that is well-equipped to efficiently serve both formal Dining Room and informal Nook

◼ Two additional bedrooms sharing a full bath on the second floor

FIRST FLOOR — 1,837 SQ. FT.
SECOND FLOOR — 848 SQ. FT.
BASEMENT — 1,803 SQ. FT.
BONUS ROOM — 288 SQ. FT.

TOTAL LIVING AREA:
2,685 SQ. FT.

Refer to **Pricing Schedule B** on the order form for pricing information

Beckoning Country Porch

■ This plan features:

— Three bedrooms

— Two full and one half baths

■ Country styled exterior with dormer windows above friendly front Porch

■ Vaulted ceiling and central fireplace accent the spacious Great Room

■ L-shaped Kitchen/Dining Room with work island and atrium door to back yard

■ First floor Master Suite with vaulted ceiling, walk-in closet, private bath and optional private Deck with hot tub

■ Two additional bedrooms on the second floor with easy access to full bath

SECOND FLOOR

Br 2
10-10 x 12-6

Br 3
11-6 x 12-6

railing

DN

open to great room below

open to master bedroom below

FIRST FLOOR — 1,061 SQ. FT.
SECOND FLOOR — 499 SQ. FT.
BASEMENT — 1,061 SQ. FT.

TOTAL LIVING AREA:
1,560 SQ. FT.

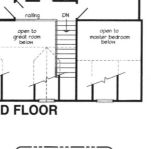

Alternate Foundation Plan

Master Br
12 x 14-6

crawl space access

furn

FIRST FLOOR
No. 34603

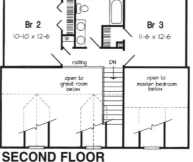

Kitchen
8-1 x 12-7

Dining
4-8 x 12-7
8' clg

Master Br
12 x 14-6
vault clg

Great Room
19-7 x 14-10
vault clg

Porch

34'-0"

40'-0"

© 1995 Donald A. Gardner Architects, Inc.

attic storage

BED RM.
10-4 x 10-0

bath

MASTER
BED RM.
13-6 x 15-8

BONUS RM.
20-0 x 14-2

cl

down

walk-in
closet

attic
storage

BED RM.
11-4 x 11-10

walk-in
closet

master
bath

SECOND FLOOR PLAN

TOTAL LIVING AREA:
1,792 SQ. FT.

Appealing Farmhouse Design

■ This plan features:

— Three bedrooms

— Two full and one half baths

■ Comfortable farmhouse features easy to build floor plan with many extras

■ Great Room which is open to the Kitchen and Breakfast bay, and expanded living space provided by the full back porch

■ For narrower lot restrictions, the Garage can be modified to open in front

■ Second floor Master Bedroom Suite contains a walk-in closet and a private bath with a garden tub and separate shower

■ Two more bedrooms on the second floor, one with a walk-in closet, share a full bath

FIRST FLOOR — 959 SQ. F.T
SECOND FLOOR — 833 SQ. FT.
BONUS ROOM — 344 SQ. FT.
GARAGE & STORAGE — 500 SQ. FT.

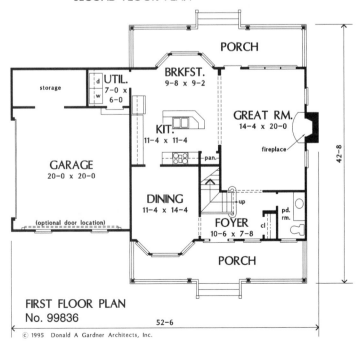

PORCH

storage

d UTIL.
w 7-0 x
6-0

BRKFST.
9-8 x 9-2

GREAT RM.
14-4 x 20-0

KIT.
11-4 x 11-4

pan.

fireplace

GARAGE
20-0 x 20-0

42-8

DINING
11-4 x 14-4

up

pd.
rm.

(optional door location)

FOYER
10-6 x 7-8

cl

PORCH

FIRST FLOOR PLAN
No. 99836

52-6

© 1995 Donald A Gardner Architects, Inc.

Refer to **Pricing Schedule D** on the order form for pricing information

Symmetrical Southern Beauty

- This plan features:

— Four bedrooms

— Three full and one half baths

- Spacious Family Room with cozy fireplace and access to Covered Porch and Patio

- Peninsula counter/eating bar and adjoining Breakfast area, Garage entry and Utility room in efficient Kitchen

- Corner Master Bedroom with large walk-in closet and double vanity bath

- First floor bedroom with a walk-in closet and private bath

- Two additional bedrooms on second floor with walk-in closets, share a full bath and balcony

- No materials list available

- Please specify a crawl space or slab foundation when ordering

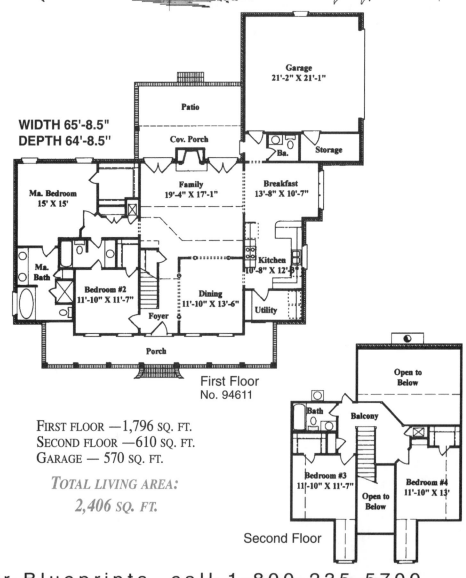

WIDTH 65'-8.5"
DEPTH 64'-8.5"

First Floor
No. 94611

FIRST FLOOR —1,796 SQ. FT.
SECOND FLOOR —610 SQ. FT.
GARAGE — 570 SQ. FT.

TOTAL LIVING AREA:
2,406 SQ. FT.

Second Floor

To order your Blueprints, call 1-800-235-5700

Refer to **Pricing Schedule C** on the order form for pricing information

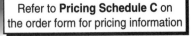

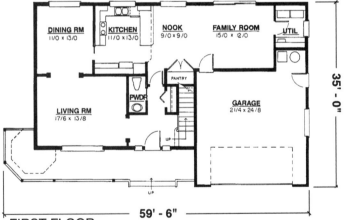

FIRST FLOOR
No. 91053

FIRST FLOOR — 1,150 SQ. FT.
SECOND FLOOR — 949 SQ. FT.
GARAGE — 484 SQ. FT.

TOTAL LIVING AREA:
2,099 SQ. FT.

SECOND FLOOR

Updated Victorian

■ This plan features:

— Three bedrooms

— Two full and one half baths

■ Classic Victorian exterior design accented by a wonderful turret room and second floor covered porch

■ Spacious formal Living Room leading into a formal Dining Room for ease in entertaining

■ Efficient, U-shaped Kitchen with loads of counter space and a peninsula snackbar, opens to an eating Nook and Family Room for informal gatherings and activities

■ Elegant Master Suite with a unique, octagon Sitting area, a private Porch, an oversized, walk-in closet and private bath with a double vanity and a window tub

■ Two additional bedrooms with ample closets share a full bath

Refer to **Pricing Schedule D** on the order form for pricing information

Quaint Charmer

■ This plan features:

— Three or four bedrooms

— Two full and one half baths

■ A vaulted ceiling in the Living Room adding to its spaciousness

■ A formal Dining Room with easy access to both the Living Room and the Kitchen

■ An efficient Kitchen with double sinks, and ample storage and counter space

■ An informal Eating Nook with a built-in pantry

■ A large Family Room with a fireplace

■ A plush Master Suite with a vaulted ceiling and luxurious Master Bath and two walk-in closets

■ Two additional bedrooms share a full bath with a convenient laundry chute

■ No materials list available

An
EXCLUSIVE DESIGN
By Energetic Enterprises

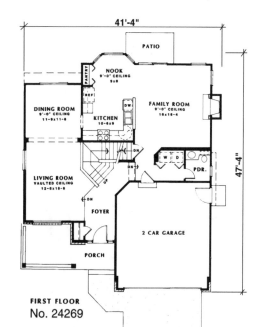

FIRST FLOOR
No. 24269

SECOND FLOOR

FIRST FLOOR — 1,115 SQ. FT.
SECOND FLOOR — 1,129 SQ. FT.
BASEMENT — 1,096 SQ. FT.
GARAGE — 415 SQ. FT.

TOTAL LIVING AREA:
2,244 SQ. FT.

To order your Blueprints, call 1-800-235-5700

PLAN NO. 93322

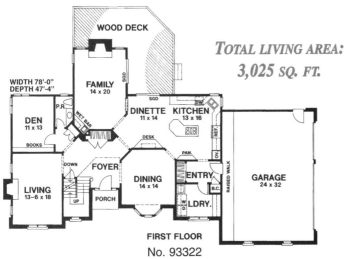

TOTAL LIVING AREA:
3,025 SQ. FT.

WIDTH 78'-0"
DEPTH 47'-4"

WOOD DECK

FAMILY
14 x 20

DEN
11 x 13

BOOKS

P.R.

WET BAR

DINETTE
11 x 14

KITCHEN
13 x 16

DW

REF.

OV.

DESK

PAN.

FOYER

DINING
14 x 14

ENTRY

B.C.

GARAGE
24 x 32

RAISED WALK

LIVING
13-6 x 18

DOWN

UP

PORCH

LDRY.
W D

FIRST FLOOR
No. 93322

ROOF

SHWR

MB

FAMILY
(Below)

RAILING

B2

T

L.C.

BR 4
12-4 x 12-6

BALCONY

RAILING

HALL

FOYER
(Below)

ROOF

MBR
13-6 x 18

DOWN

BR 2
14 x 12

BR 3
12 x 12-6

SECOND FLOOR

An
EXCLUSIVE DESIGN
By Patrick Morabito, A.I.A. Architect

Splendor and Distinction

■ This plan features:

— Four bedrooms

— Two full and one half baths

■ An expansive Kitchen with a cooktop island/eating bar and corner double sinks

■ A spacious Family Room equipped with a built-in wetbar and a cozy fireplace

■ A formal Living Room with a second fireplace

■ A bay window adding elegance to the formal Dining Room

■ A pampering Master Bath adding to the privacy of the Master Suite

■ Three additional bedrooms that share a full hall bath

■ A balcony that overlooks the Foyer and the Family Room

■ No materials list available

FIRST FLOOR — 1,720 SQ. FT.
SECOND FLOOR — 1,305 SQ. FT.
BASEMENT — 1,720 SQ. FT.
GARAGE — 768 SQ. FT.

Refer to **Pricing Schedule B** on the order form for pricing information

Reminiscent of an Earlier Time

- This plan features:

—Three bedrooms

—Two full and one half baths

- Spacious Great Room accented by high windows to either side of the cozy fireplace

- A grand opening to the Dining Room from the Great Room allowing this area to be visually open

- Efficient Kitchen serves the Dining Room with ease

- Small windows across the second floor provide light to the plant shelf, stairway and balcony as well as light and privacy to the Master Bath

- Master Bedroom suite located close to the secondary bedrooms providing a sense of security for the younger family members

- Bonus room above the garage offers expandable space for your enjoyment

SECOND FLOOR

Bonus Room 14'5" x 11'6"
Bedroom 10'8" x 10'4"
Bedroom 10'10" x 10'8"
Hall
Bath
Open
Master Bedroom 12' x 16'
Bath
walk-in closet
walk-in closet
plant shelf

- No materials list is available for this plan

FIRST FLOOR — 733 SQ. FT.
SECOND FLOOR — 819 SQ. FT.
BONUS ROOM — 236 SQ. FT.

TOTAL LIVING AREA:
1,552 SQ. FT.

Two-car Garage 19'6" x 23'8"
Bath
Kitchen 10'5" x 13'5"
Dining Room 12' x 12'8"
Great Room 20' x 14'6"
Foyer
Porch
FIRST FLOOR No. 92686
37'6"
46'8"

To order your Blueprints, call 1-800-235-5700

© 1993 Donald A. Gardner Architects, Inc.

B. NATHAN

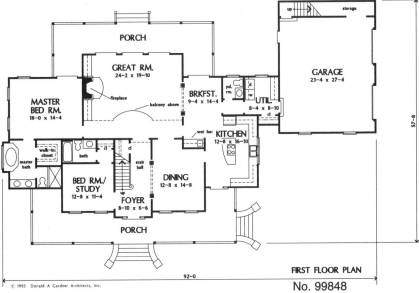

FIRST FLOOR PLAN

No. 99848

© 1993 Donald A Gardner Architects, Inc.

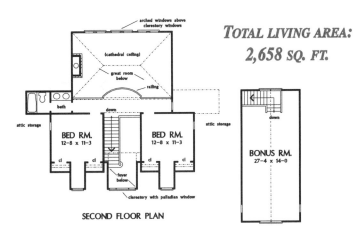

SECOND FLOOR PLAN

TOTAL LIVING AREA: 2,658 SQ. FT.

Rambling Farmhouse

■ This plan features:

— Four bedrooms

— Three full and one half baths

■ Two-story Foyer with palladium window above

■ Great Room topped by a vaulted ceiling and highlighted by a fireplace and a balcony overlooking the room

■ Great Room, Breakfast Room, and Master Bedroom access Porch for open circulation

■ Nine foot ceilings throughout the first floor

■ Island Kitchen with direct access to formal and informal eating areas

■ First floor Study/bedroom with private full bath

FIRST FLOOR — 2,064 SQ. FT.
SECOND FLOOR — 594 SQ. FT.
GARAGE & STORAGE — 710 SQ. FT.
BONUS ROOM — 483 SQ. FT.

Refer to **Pricing Schedule E** on the order form for pricing information

An
EXCLUSIVE DESIGN
By Patrick Morabito, A.I.A. Architect

With Room for All

◼ This plan features:

— Four bedrooms

— Two full and one half baths

◼ A formal Dining and Living Room to either side of the Foyer

◼ Pocket doors separating the Family Room from the Living Room

◼ A cooktop island Kitchen with a corner pantry and ample counter and storage space

◼ A open layout between the Kitchen, Dinette and the Family Room creating a spacious family living area

◼ A secluded Den, equipped with built-in shelves and a sunny bay window

◼ A Master Suite with a lavish compartmented Bath and a walk-in closet

◼ No materials list is available for this plan

FIRST FLOOR — 1,536 SQ. FT.
SECOND FLOOR — 1,245 SQ. FT.

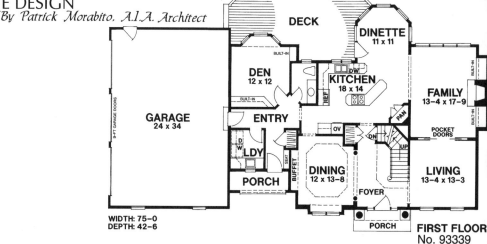

WIDTH: 75–0
DEPTH: 42–6

FIRST FLOOR
No. 93339

TOTAL LIVING AREA:
2,781 SQ. FT.

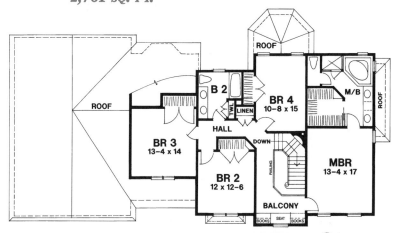

SECOND FLOOR

To order your Blueprints, call 1-800-235-5700

Unique Tower

■ This plan features:

— Three bedrooms

— Two full and one half baths

■ Extensive porch wraps around and accesses active areas of home

■ Combined Living and Dining area offers comfortable entertaining

■ Efficient Kitchen with a cooktop work island, built-in desk, nearby laundry and Garage entry

■ An expansive Family Room with a cozy fireplace and atrium door to Porch

■ Unique Den with loads of light and built-in shelves

■ Vaulted ceiling, walk-in closet and a luxurious bath enhance Master Suite

■ Two additional bedrooms, one with a unique shape, share a full bath

FIRST FLOOR — 1,337 SQ. FT.
SECOND FLOOR — 1,025 SQ. FT.

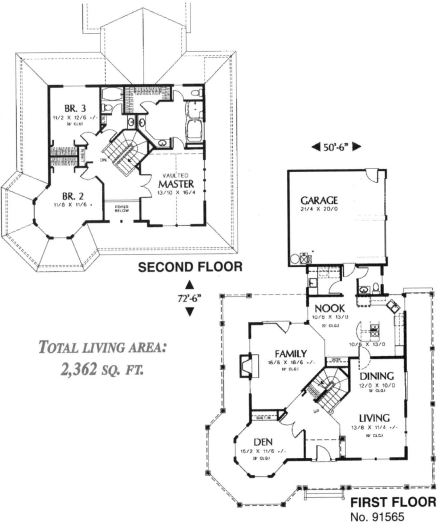

SECOND FLOOR

TOTAL LIVING AREA:
2,362 SQ. FT.

◄ 50'-6" ►

72'-6"

FIRST FLOOR
No. 91565

Refer to **Pricing Schedule E** on the order form for pricing information

Relax on the Veranda

■ This plan features:

— Four bedrooms

— Three full and one half baths

■ A wrap-around veranda

■ A skylit Master Suite with elevated custom spa, twin basins, a walk-in closet, and an additional vanity outside the bathroom

■ A vaulted ceiling in the Den

■ A fireplace in both the Family Room and the formal Living Room

■ An efficient Kitchen with a peninsula counter and a double sink

■ Two additional bedrooms with walk-in closets are served by a compartmentalized bath

■ A Guest Suite with a private bath

MAIN AREA — 3,051 SQ. FT.
GARAGE — 646 SQ. FT.

TOTAL LIVING AREA:
3,051 SQ. FT.

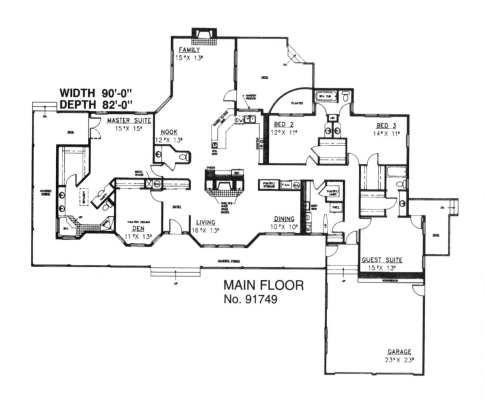

WIDTH 90'-0"
DEPTH 82'-0"

MAIN FLOOR
No. 91749

To order your Blueprints, call 1-800-235-5700

B. NATHAN

©1995 Donald A. Gardner Architects, Inc.

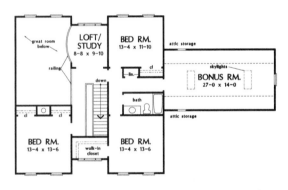

LOFT/ STUDY
8–8 x 9–10

great room below

railing

BED RM.
13–4 x 11–10

attic storage

skylights

down

BONUS RM.
27–0 x 14–0

lin.

cl

bath

attic storage

BED RM.
13–4 x 13–6

walk–in closet

BED RM.
13–4 x 13–6

cl   cl

SECOND FLOOR PLAN

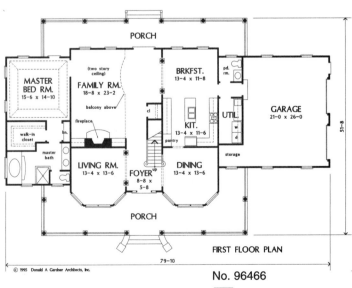

PORCH

MASTER BED RM.
15–6 x 14–10

FAMILY RM.
18–8 x 23–2
(two story ceiling)

BRKFST.
13–4 x 11–8

pd. rm.

balcony above

GARAGE
21–0 x 26–0

fireplace

walk–in closet

lin.

cl

KIT.
13–4 x 11–6

UTIL.

master bath

pantry

storage

51–8

LIVING RM.
13–4 x 13–6

FOYER
8–8 x 5–8

up

DINING
13–4 x 13–6

PORCH

© 1995 Donald A Gardner Architects, Inc.

79–10

FIRST FLOOR PLAN

No. 96466

With Room to Grow

■ This plan features:

— Four bedrooms

— Two full and one half baths

■ Symmetrical gables, large bay windows, and expansive porches welcome your family home

■ A voluminous Family Room with a two story ceiling and built-in cabinets on either side of the fireplace

■ Master Suite is highlighted by a tray ceiling, ample closet space, and a lush, private bath

■ Three spacious bedrooms share a full bath and easy access to the bonus room highlighted by skylights

FIRST FLOOR — 1,943 SQ. FT.
SECOND FLOOR — 1,000 SQ. FT.
BONUS AREA — 403 SQ. FT.
GARAGE — 601 SQ. FT.

TOTAL LIVING AREA:
2,943 SQ. FT.

Refer to **Pricing Schedule D** on the order form for pricing information

Cottage Influence

■ This plan features:

— Three or four bedrooms

— Three full and one half baths

■ Cozy porch entrance into Foyer with banister staircase and coat closet

■ Expansive Great Room has focal point fireplace and access to Covered Porch

■ Cooktop island in Kitchen easily serves Breakfast bay and formal Dining Room

■ Large Master Suite has access to Covered Porch

■ Study/Guest Bedroom with private access to a full bath

■ Two second floor bedrooms with dormers, private vanities and walk-in closets

■ No materials list available

■ Please specify a crawl space or slab foundation when ordering

FIRST FLOOR — 1,916 SQ. FT.
SECOND FLOOR — 617 SQ. FT.
GARAGE — 516 SQ. FT.

TOTAL LIVING AREA:
2,533 SQ. FT.

Width — 66'-0"
Depth — 66'-0"

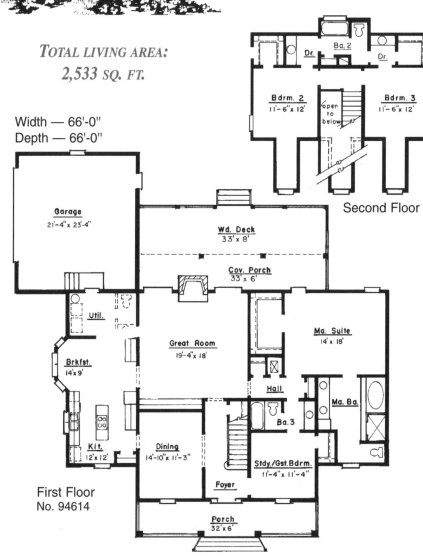

Second Floor

Bdrm. 2 — 11'-6" x 12'
Bdrm. 3 — 11'-6" x 12'
Ba. 2
Dr.
Dr.
open to below

First Floor
No. 94614

Garage — 21'-4" x 23'-4"
Wd. Deck — 33' x 8'
Cov. Porch — 33' x 6'
Util.
Brkfst. — 14' x 9'
Great Room — 19'-4" x 18'
Ma. Suite — 14' x 18'
Hall
Ma. Ba.
Ba. 3
Kit. — 12' x 12'
Dining — 14'-10" x 11'-3"
Foyer
Stdy./Gst.Bdrm. — 11'-4" x 11'-4"
Porch — 32' x 6'

To order your Blueprints, call 1-800-235-5700

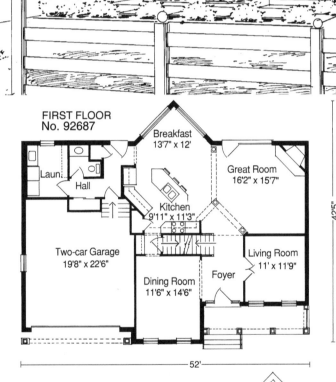

PLAN NO. 92687

FIRST FLOOR
No. 92687

Breakfast
13'7" x 12'

Great Room
16'2" x 15'7"

Laun.

Hall

Kitchen
9'11" x 11'3"

Living Room
11' x 11'9"

Two-car Garage
19'8" x 22'6"

Dining Room
11'6" x 14'6"

Foyer

42'5"

52'

SECOND FLOOR

Bedroom
11' x 11'8"

Master Bedroom
14'1" x 14'1"

Bonus Room
17'8" x 15'

Balcony
wood rail

Bath

stairs brf.

slope ceiling

Bedroom
11'6" x 12'1"

FIRST FLOOR — 1,218 SQ. FT.
SECOND FLOOR — 904 SQ. FT.
BONUS ROOM — 328 SQ. FT.

TOTAL LIVING AREA:
2,122 SQ. FT.

Warm and Friendly Atmosphere

▪ This plan features:

—Three bedrooms

—Two full and one half baths

▪ Exterior comprised of arched detailing, wood trim and a front porch sets the stage for a warm and friendly atmosphere

▪ Formal and informal interior spaces create usable rooms for all occasions

▪ Angled island defines the Kitchen and visually opens the Breakfast and Great Room

▪ Second floor balcony overlooks the Foyer and leads to the Master Bedroom

▪ Master Suite topped by a tray ceiling and highlighted by a spacious bath creating a comfortable retreat

▪ Bonus room offers the option of expanding to fit your needs

▪ No materials list is available for this plan

Conventional and Classic Comfort

■ This plan features:

— Three bedrooms

— Two full and one half baths

■ Porch accesses two-story Foyer with decorative window and a landing staircase

■ Formal Dining Room accented by a recessed window

■ Spacious Family Room crowned by a vaulted ceiling over a hearth fireplace

■ Efficient Kitchen with an extended counter and bright Dinette area with bay window

■ Convenient Laundry, Powder Room and Garage entry near Kitchen

■ First floor Master Bedroom has a walk-in closet and Master Bath

■ Two additional bedrooms and a full bath complete second floor

■ No materials list available

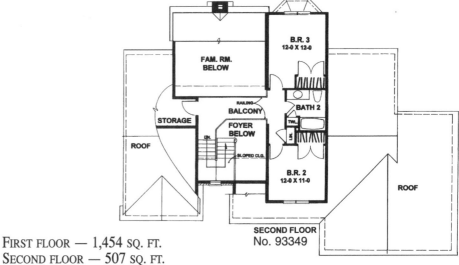

SECOND FLOOR
No. 93349

FIRST FLOOR — 1,454 SQ. FT.
SECOND FLOOR — 507 SQ. FT.
BASEMENT — 1,454 SQ. FT.
GARAGE — 624 SQ. FT.

TOTAL LIVING AREA:
1,961 SQ. FT.

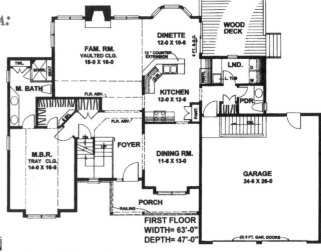

FIRST FLOOR
WIDTH= 63'-0"
DEPTH= 47'-0"

An
EXCLUSIVE DESIGN
By Patrick Morabito, A.I.A. Architect

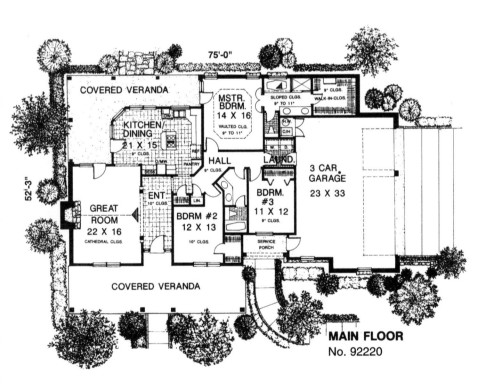

75'-0"

52'-3"

COVERED VERANDA

KITCHEN/
DINING
21 X 15
9" CLGS.

MSTR.
BDRM.
14 X 16
VAULTED CLG.
9" TO 11"

SLOPED CLGS.
9" TO 11"

9" CLGS.

WALK-IN-CLOS.

HALL
9" CLGS.

PANTRY

REF.

LAUND.

W. D.

C/H

3 CAR
GARAGE
23 X 33

ENT.
10" CLGS.

LIN.

BDRM.
#3
11 X 12
9" CLGS.

GREAT
ROOM
22 X 16
CATHEDRAL CLGS.

BDRM #2
12 X 13
10" CLGS.

SERVICE
PORCH

COVERED VERANDA

MAIN FLOOR
No. 92220

TOTAL LIVING AREA:
1,830 SQ. FT.

Southern Hospitality

■ This plan features:

— Three bedrooms

— Two full baths

■ Welcoming Covered Veranda

■ Easy-care, tiled Entry leads into Great Room with fieldstone fireplace and atrium door to another Covered Veranda topped by a cathedral ceiling

■ A bright Kitchen/Dining Room includes a stovetop island/snackbar, built-in pantry and desk and access to Covered Veranda

■ Vaulted ceiling crowns Master Bedroom that offers a plush bath and huge walk-in closet

■ Two additional bedrooms with ample closets share a double vanity bath

■ No materials list is available for this plan.

MAIN FLOOR — 1,830 SQ. FT.
GARAGE — 759 SQ. FT.

Refer to **Pricing Schedule E** on the order form for pricing information

Classic Front Porch

■ This plan features:

— Four bedrooms

— Two full and one half baths

■ Stone and columns accenting the wrap-around front porch

■ A formal Living Room and Dining Room adjoining with columns at their entrances

■ An island Kitchen with a double sink, plenty of cabinet and counter space and a walk-in pantry

■ A Breakfast Room flowing into the Family Room and the Kitchen

■ A corner fireplace and a built-in entertainment center in the Family Room

■ Lavish Master Suite with a decorative ceiling and an ultra bath

■ Three roomy, additional bedrooms sharing a full hall bath

FIRST FLOOR — 1,584 SQ. FT.
SECOND FLOOR — 1,277 SQ. FT.
GARAGE — 550 SQ. FT.
BASEMENT — 1,584 SQ. FT.

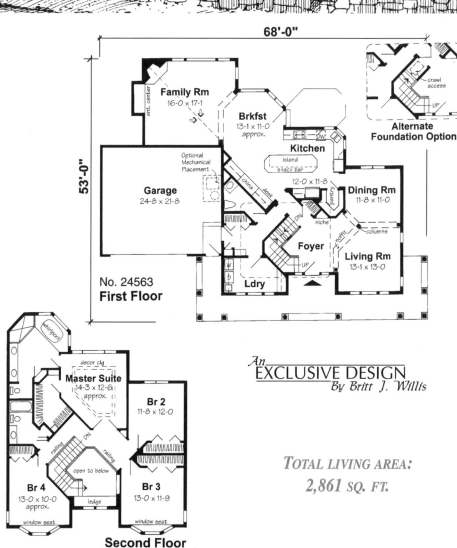

An
EXCLUSIVE DESIGN
By Britt J. Willis

TOTAL LIVING AREA:
2,861 SQ. FT.

To order your Blueprints, call 1-800-235-5700

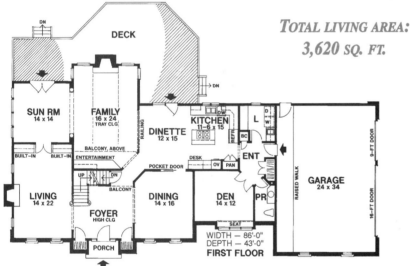

TOTAL LIVING AREA:
3,620 SQ. FT.

DECK

SUN RM
14 x 14

FAMILY
16 x 24
TRAY CLG.

KITCHEN
11-6 x 15

DINETTE
12 x 15

BALCONY, ABOVE

RAILING

ENTERTAINMENT

BUILT-IN BUILT-IN

DESK

D
DW
L
REFR.
BC

D
W
OV PAN

ENT

LIVING
14 x 22

UP DN

BALCONY

POCKET DOOR

DINING
14 x 16

DEN
14 x 12

PR

GARAGE
24 x 34

RAISED WALK

9-FT DOOR

16-FT DOOR

FOYER
HIGH CLG

SEAT

WIDTH — 86'-0"
DEPTH — 43'-0"
FIRST FLOOR

PORCH

An
EXCLUSIVE DESIGN
By Patrick Morabito, A.I.A. Architect

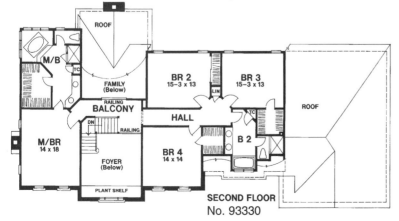

ROOF

M/B

TC

FAMILY
(Below)

BR 2
15-3 x 13

BR 3
15-3 x 13

LIN

RAILING
BALCONY

HALL

ROOF

M/BR
14 x 18

DN RAILING

FOYER
(Below)

BR 4
14 x 14

TC

B 2

PLANT SHELF

SECOND FLOOR
No. 93330

A Grand Presence

◼ This plan features:

— Four bedrooms

— Two full and one half baths

◼ Gourmet Kitchen with a cooktop island and built-in pantry

◼ Formal Living Room with a fireplace

◼ Pocket doors separate the formal Dining Room from the Dinette area

◼ A balcony overlooks the Family Room

◼ Expansive Family Room with a fireplace and a built-in entertainment center

◼ A luxurious Master Bath that highlights the Master Suite

◼ Three additional bedrooms that share use of a compartmented full hall bath

◼ No materials list available

FIRST FLOOR — 2,093 SQ. FT.
SECOND FLOOR — 1,527 SQ. FT.
BASEMENT — 2,093 SQ. FT.
GARAGE — 816 SQ. FT.

Refer to **Pricing Schedule E** on the order form for pricing information

Multiple Porches Provide Added Interest

◼ This plan features:

— Four bedrooms

— Three full and one half baths

◼ Two-story central Foyer flanked by Living and Dining rooms

◼ Spacious Great Room with large fireplace between french doors to Porch and Deck

◼ Country-size Kitchen with cooktop work island, walk-in pantry and Breakfast area with Porch access

◼ Pampering Master Bedroom offers a decorative ceiling, sitting area, Porch and Deck access, a huge walk-in closet and lavish bath

◼ Three second floor bedrooms with walk-in closets, have private access to a full bath

◼ No materials list is available for this plan

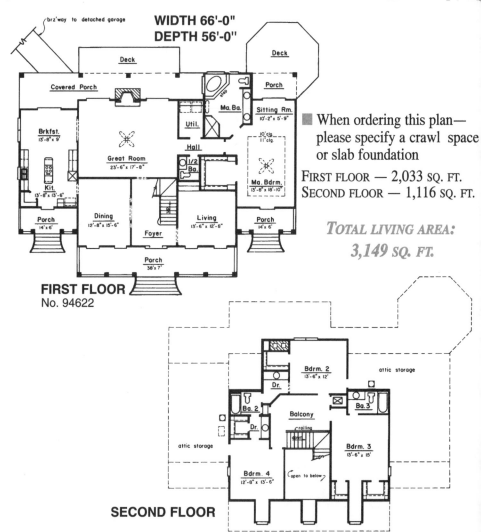

WIDTH 66'-0"
DEPTH 56'-0"

brz'way to detached garage

Deck

Deck

Covered Porch

Porch

Ma. Ba.

Sitting Rm.
10'-2" x 5'-9"

Brkfst.
13'-8" x 9'

Util.

10' clg
11' clg

Hall

Great Room
23'-6" x 17'-8"

1/2 Ba.

Kit.
13'-8" x 13'-6"

Ma. Bdrm.
13'-8" x 18'-10"

Porch
14' x 6'

Dining
12'-8" x 15'-6"

Living
13'-6" x 12'-8"

Porch
14' x 6'

Foyer

Porch
38' x 7'

FIRST FLOOR
No. 94622

◼ When ordering this plan— please specify a crawl space or slab foundation

FIRST FLOOR — 2,033 SQ. FT.
SECOND FLOOR — 1,116 SQ. FT.

TOTAL LIVING AREA: 3,149 SQ. FT.

Bdrm. 2
13'-6" x 12'

attic storage

Dr.

Ba. 2

Ba. 3

Dr.

Balcony
railing

attic storage

Bdrm. 3
13'-6" x 15'

attic storage

Bdrm. 4
12'-8" x 13'-6"

open to below

SECOND FLOOR

To order your Blueprints, call 1-800-235-5700

TOTAL LIVING AREA:
2,218 SQ. FT.

MAIN FLOOR
No. 90454

Traditional Ranch Plan

◼ This plan features:

—Three bedrooms

—Two full baths

◼ Large Foyer set between the formal Living and Dining Rooms

◼ Spacious Great Room adjacent to the open Kitchen /Breakfast area

◼ Secluded Master Bedroom highlighted by the Master Bath with a garden tub, separate shower, and his and her vanity

◼ Bay window allows bountiful natural light into the Breakfast area

◼ Two additional bedrooms sharing a full bath

◼ When ordering this plan—please specify basement or crawl space foundation

MAIN FLOOR — 2,218 SQ. FT.
BASEMENT — 1,658 SQ. FT.
GARAGE — 528 SQ. FT.

Refer to **Pricing Schedule E** on the order form for pricing information

Elegant and Inviting

■ This plan features:

— Five bedrooms

— Three and one half baths

■ Wrap-around verandas and a three-season porch

■ An elegant Parlor with a parquet floor and a formal Dining Room separated by a half-wall

■ An adjoining Kitchen with a Breakfast bar and nook

■ A Gathering Room with a fireplace, soaring ceilings and access to the porch

FIRST FLOOR — 1,580 SQ. FT.
SECOND FLOOR — 1,164 SQ. FT.
BASEMENT — 1,329 SQ. FT.
GARAGE — 576 SQ. FT.

TOTAL LIVING AREA:
2,744 SQ. FT.

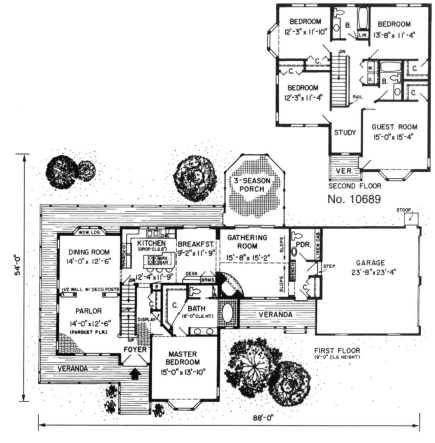

No. 10689

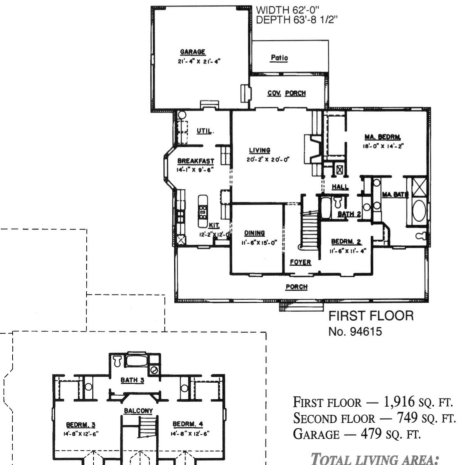

WIDTH 62'-0"
DEPTH 63'-8 1/2"

FIRST FLOOR
No. 94615

SECOND FLOOR

FIRST FLOOR — 1,916 SQ. FT.
SECOND FLOOR — 749 SQ. FT.
GARAGE — 479 SQ. FT.

TOTAL LIVING AREA:
2,665 SQ. FT.

Grand Country Porch

■ This plan features:

— Four bedrooms

— Three full baths

■ Large front Porch provides shade and Southern hospitality

■ Spacious Living Room has access to Covered Porch and Patio, and a cozy fireplace between built-in shelves

■ Country Kitchen with a cooktop island, bright Breakfast bay, Utility Room, Garage entry, and adjoining Dining Room

■ Corner Master Bedroom has a walk-in closet and private bath

■ First floor bedroom with private access to a full bath

■ Two additional second floor bedrooms with dormers, walk-in closets and vanities, share a full bath

■ No materials list available

■ Please specify a crawl space or slab foundation when ordering

© 1997 Donald A Gardner Architects, Inc.

Country Style Home With Corner Porch

■ This plan features:

— Three bedrooms

— Two full baths

■ Dining Room has four floor to ceiling windows that overlook front porch

■ Great Room topped by a cathedral ceiling, enhanced by a fireplace, and sliding doors to the back porch

■ Utility Room located near Kitchen and Breakfast Nook

■ Master Bedroom has a walk in closet and private bath

■ Two additional bedrooms with ample closet space share a full bath

■ A skylight bonus room over the two car garage

MAIN FLOOR — 1,815 SQ. FT.
GARAGE — 522 SQ, FT,

TOTAL LIVING AREA:
1,815 SQ. FT.

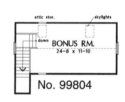

No. 99804

FLOOR PLAN

To order your Blueprints, call 1-800-235-5700

No. 20136
Second Floor

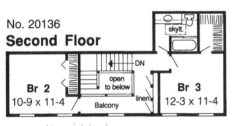

Br 2
10-9 x 11-4

open to below

DN

linen

Balcony

Br 3
12-3 x 11-4

skylt.

FIRST FLOOR — 1,556 SQ. FT.
SECOND FLOOR — 539 SQ. FT.
GARAGE — 572 SQ. FT.
BASEMENT — 1,556 SQ. FT.

TOTAL LIVING AREA:
2,095 SQ. FT.

Colonial Classic

◼ This plan features:

— Three bedrooms

— Two full and one half bath

◼ A colonial exterior with an modern interior provides charm and comfort

◼ Center Foyer with a lovely, landing staircase and balcony, flanked by Formal Parlor and Dining Room

◼ Expansive Living Room with a cozy fireplace below decorative beams on sloped ceiling

◼ Hub Kitchen with pantry and peninsula counter, easily serves Breakfast area, Dining Room

◼ Private Master Bedroom offers a decorative ceiling, over-sized closet and double vanity bath

◼ Two second floor bedrooms with ample closets, share a full bath

7-1/2" ceiling reveal

MBr 1
14-10 x 14

lin.

Brkfst
10-8 x 8

Deck

sloped ceiling

Living Rm
24 x 13-6

decor. beams

Kit
11 x 13-4

W D

Ldry

Garage
21-8 x 25-6

pan.

UP

open to above

DN

Parlor
13-2 x 11-4

Dining
13-6 x 11-4

DN

1-1/2" ceiling reveal

Foyer

First Floor

46'-0"

64'-0"

An
EXCLUSIVE DESIGN
By Karl Kreeger

Refer to **Pricing Schedule C** on the order form for pricing information

© 1991 Donald A. Gardner Architects, Inc.

Country Farmhouse

■ This plan features:

— Three bedrooms

— Two full and one half baths

■ Palladian window in clerestory dormer bathes two-story Foyer in natural light

■ Private Master Bedroom Suite offers everything: walk-in closet, whirlpool tub, shower, and double vanity

■ Two bedrooms upstairs with dormers and storage access share a full bath

■ Skylit Bonus Room over Garage and optional basement

■ Please specify a basement or crawl space option when ordering

FIRST FLOOR — 1,356 SQ. FT.
SECOND FLOOR — 542 SQ. FT.
BONUS ROOM — 393 SQ. FT.
GARAGE & STORAGE — 543 SQ. FT.

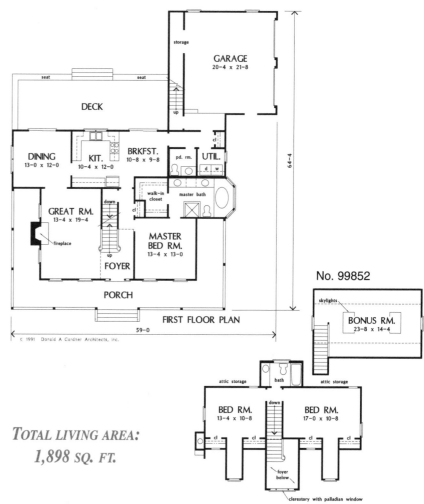

TOTAL LIVING AREA:
1,898 SQ. FT.

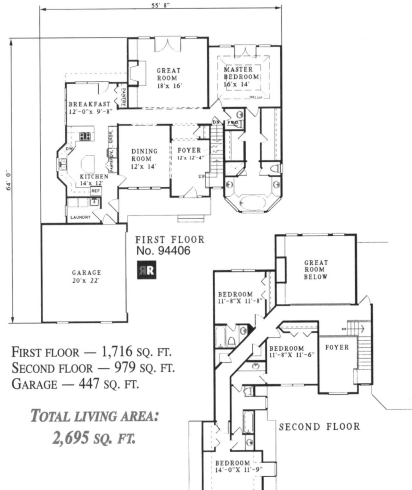

FIRST FLOOR
No. 94406

GREAT ROOM 18'x 16'

MASTER BEDROOM 16'x 14'

BREAKFAST 12'-0"x 9'-8"

DINING ROOM 12'x 14'

FOYER 12'x 12'-4'

KITCHEN 14'x 12'

LAUNDRY

GARAGE 20'x 22'

55' 8"

64' 0"

GREAT ROOM BELOW

BEDROOM 11'-8"X 11'-8"

BEDROOM 11'-8"X 11'-6"

FOYER

BEDROOM 14'-0"X 11'-9"

SECOND FLOOR

FIRST FLOOR — 1,716 SQ. FT.
SECOND FLOOR — 979 SQ. FT.
GARAGE — 447 SQ. FT.

TOTAL LIVING AREA:
2,695 SQ. FT.

Elegant European Style

■ This plan features:

— Four bedrooms

— Three full and one half baths

■ Distinctive two-story Foyer with detailed windows and a lovely landing staircase

■ Formal Dining Room highlighted by a decorative window

■ Spacious Great Room enhanced by a huge fireplace between built-ins and French doors to outdoors

■ Convenient Kitchen with two pantries, a desk, extended serving counter, bright Breakfast area, and nearby laundry/Garage entry

■ Secluded Master Bedroom offers a trey ceiling, outdoor access, two walk-in closets and a lavish bath

■ No materials list is available for this plan

■ This plan is available with a basement or crawl space foundation. Please specify when ordering.

©1994 Donald A. Gardner Architects, Inc.

Refer to **Pricing Schedule C** on the order form for pricing information

Maximum Use of Space

■ This plan features:

— Three bedrooms

— Two full and one half baths

■ A smart angled Kitchen opening to the large Great Room/Dining area forming an efficient, L-shaped living space

■ Operable skylights in the vaulted ceiling visually expand the Great Room

■ Deep front and back porches extend living to the outdoors

■ Master Bedroom with access to the back porch, a walk-in closet, and a luxurious bath with whirlpool tub, separate shower and dual vanity

■ Two bedrooms and a full bath on the second floor

FIRST FLOOR — 1,164 SQ. FT.
SECOND FLOOR — 458 SQ. FT.
GARAGE & STORAGE — 482 SQ. FT.

TOTAL LIVING AREA:
1,622 SQ. FT.

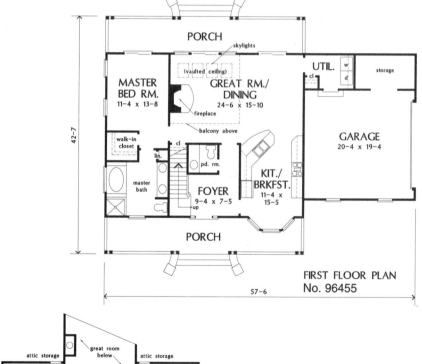

FIRST FLOOR PLAN
No. 96455

SECOND FLOOR PLAN

To order your Blueprints, call 1-800-235-5700

SECOND FLOOR

ROOF

BED RM 11'-0" x 10'-0"

BATH

BED RM 13'-4" x 11'-0"

skylights

cl

lin

dn

cl

stor.

H

railing

lin

BATH

DECK

W.I.C.

MASTER SUITE 15'-4" x 12'-8"

high ceiling

railing

TOWER

ROOF

ROOF

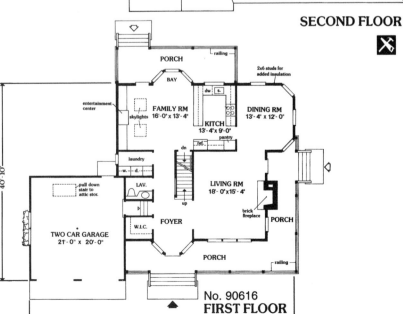

Victorian Touches Disguise Modern Design

■ This plan features:

— Three bedrooms

— Two full and one half baths

■ A Master Suite with a high ceiling, an arched window, a private bath, and a tower sitting room with an adjoining roof deck

■ Two additional bedrooms that share a full hall bath

■ A Living Room accentuated by a brick fireplace

■ A well-equipped Kitchen with a built-in pantry and peninsula counter

■ A sky-lit Family Room with a built-in entertainment center

FIRST FLOOR — 1,146 SQ. FT.
SECOND FLOOR — 846 SQ. FT.
BASEMENT — 967 SQ. FT.
GARAGE — 447 SQ. FT.

TOTAL LIVING AREA:
1,992 SQ. FT.

FIRST FLOOR

PORCH

railing

BAY

2x6 studs for added insulation

entertainment center

FAMILY RM 16'-0" x 13'-4"

skylights

DINING RM 13'-4" x 12'-0"

dw

s.

KITCH 13'-4" x 9'-0"

pantry

laundry

dn

ref.

w. d.

LIVING RM 18'-0" x 15'-4"

LAV.

pull down stair to attic stor.

up

brick fireplace

PORCH

TWO CAR GARAGE 21'-0" x 20'-0"

W.I.C.

FOYER

PORCH

railing

No. 90616
FIRST FLOOR

40'-0"

57'-0"

Refer to **Pricing Schedule B** on the order form for pricing information

Convenient Country

■ This plan features:

— Three bedrooms

— Two and a half baths

■ Full front Porch provides comfortable visiting and a sheltered entrance

■ Expansive Living Room with an inviting fireplace opens to bright Dining Room and Kitchen

■ U-shaped Kitchen with peninsula serving counter, Dining Room and nearby Pantry, Laundry and Garage entry

■ Secluded Master Bedroom with two closets and a double vanity bath

■ Two second floor bedrooms with ample closets and dormer windows, share a full bath

■ No materials list available

FIRST FLOOR — 1,108 SQ. FT.
SECOND FLOOR — 659 SQ. FT.
BASEMENT — 875 SQ. FT.

SECOND FLOOR

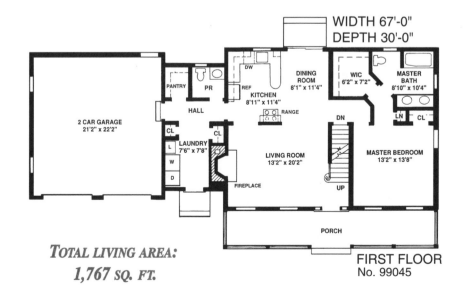

WIDTH 67'-0"
DEPTH 30'-0"

FIRST FLOOR
No. 99045

TOTAL LIVING AREA:
1,767 SQ. FT.

To order your Blueprints, call 1-800-235-5700

Refer to **Pricing Schedule F** on the order form for pricing information

Upper Floor

Bed#4
13x11

Sloping Clg.

Balcony
10'Clg.
DN

Bed#3
13x14

Ent Below
21'Clg.

Bed#2
15x11

FIRST FLOOR — 2,432 SQ. FT.
SECOND FLOOR — 903 SQ. FT.
BASEMENT — 2,432 SQ. FT.

TOTAL LIVING AREA:
3,335 SQ. FT.

Columned Entrance

■ This plan features:

— Four bedrooms

— Three and one half baths

■ Entry hall with a graceful landing staircase, flanked by formal Living and Dining rooms

■ Fireplaces highlight both the Living Room/Parlor and formal Dining Room

■ An efficient Kitchen with an island cooktop, built-in pantry and open Breakfast area

■ Cathedral ceiling crowns expansive Family Room, accented by a fireplace and a built-in entertainment center

■ Lavish Master Bedroom wing with plenty of storage space

■ Three additional bedrooms, one with a private bath, on second floor

■ No materials list available

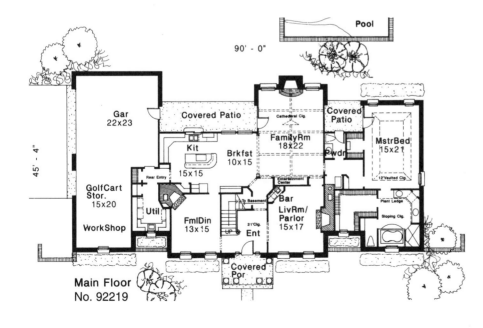

Pool

90' - 0"

45' - 4"

Gar
22x23

Covered Patio

Cathedral Clg.

Covered
Patio

FamilyRm
18x22

MstrBed
15x21

Kit

Brkfst
10x15
15x15

Rear Entry

Powdr

12'Vaulted Clg.

GolfCart
Stor.
15x20

Entertainment
Center

Plant Ledge

Util

To Basement

Bar
LivRm/
Parlor
15x17

Sloping Clg.

WorkShop

FmlDin
13x15

21'Clg.
UP
Ent

Main Floor
No. 92219

Covered
Por

Refer to **Pricing Schedule B** on the order form for pricing information

Cozy Three Bedroom

■ This plan features:

— Three bedrooms

— Two full baths

■ Covered entry leads into Activity Room highlighted by a double window and a vaulted ceiling

■ Efficient Kitchen with work island, nearby laundry and Garage entry, opens to Dining area with access to Sun Deck

■ Plush Master Bed Room offers a decorative ceiling, walk-in closet and whirlpool tub

■ Two additional bedrooms, one with a vaulted ceiling, share a full bath

■ Garage with entry into Laundry Room serving as a Mud Room

■ When ordering this plan—please specify a basement, slab or crawl space foundation

MAIN FLOOR — 1,199 SQ. FT.
GARAGE — 287 SQ. FT.

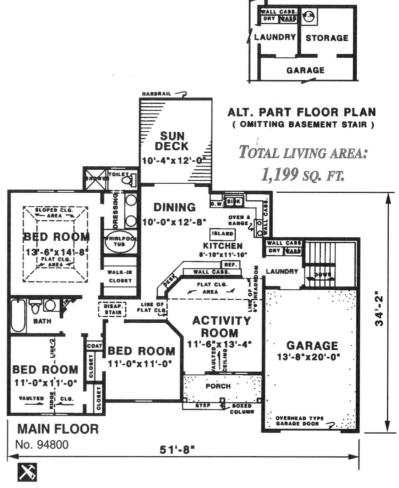

ALT. PART FLOOR PLAN
(OMITTING BASEMENT STAIR)

TOTAL LIVING AREA:
1,199 SQ. FT.

MAIN FLOOR
No. 94800

To order your Blueprints, call 1-800-235-5700

1991 Donald A. Gardner Architects, Inc.

Compact Country Cottage

- ■ This plan features:
- — Three bedrooms
- — Two full baths
- ■ Foyer opening to a large Great Room with a fireplace and a cathedral ceiling
- ■ Efficient U-shaped Kitchen with peninsula counter extending work space and separating it from the Dining Room
- ■ Two front bedrooms, one with a bay window, the other with a walk-in closet, sharing a full bath in the hall
- ■ Master Suite located to the rear with a walk-in closet and a private bath with a dual vanity
- ■ Partially covered Deck with sky-lights is accessible from Dining Room, Great Room and Master Bedroom

MAIN FLOOR — 1,310 SQ. FT.
GARAGE & STORAGE — 455 SQ. FT.

TOTAL LIVING AREA:
1,310 SQ. FT.

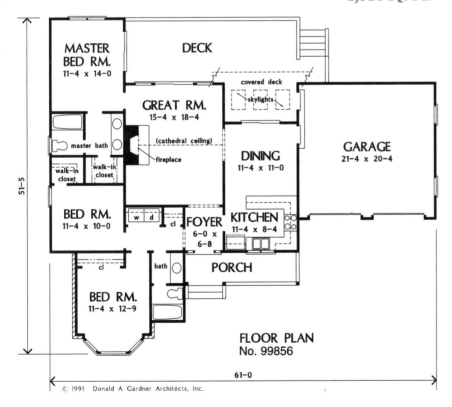

FLOOR PLAN
No. 99856

© 1991 Donald A Gardner Architects, Inc.

Refer to **Pricing Schedule A** on the order form for pricing information

Home Sweet Home

■ This plan features:

— Three bedrooms

— Two full baths

■ Single level format allows for step-saving convenience

■ Large Living Room, highlighted by a fireplace and built-in entertainment center, adjoins the Dining Room

■ Skylights, a ceiling fan and room defining columns accent the Dining Room

■ A serving bar to the Dining Room, and ample counter and cabinet space in the Kitchen

■ Decorative ceiling treatment over the Master Bedroom and a private Master Bath

■ Two secondary bedrooms with easy access to the full bath in the hall

■ No materials list available

MAIN FLOOR — 1,112 SQ. FT.
GARAGE — 563 SQ. FT.

MAIN FLOOR
No. 24723

TOTAL LIVING AREA:
1,112 SQ. FT.

An EXCLUSIVE DESIGN
By Patrick Morabito, A.I.A. Architect

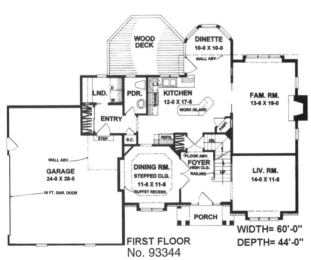

FIRST FLOOR
No. 93344

WIDTH= 60'-0"
DEPTH= 44'-0"

TOTAL LIVING AREA:
2,259 SQ. FT.

SECOND FLOOR

Elegant Dining Room

■ This plan features:

— Four Bedrooms

—Two full and one half bath

■ Two-story Foyer with graceful staircase has access to the Living and Dining rooms

■ Double door entry into Dining Room with a boxed bay window, stepped ceiling and buffet recess

■ Work island/snack bar, corner pantry and a bright Dinette bay with Deck access

■ Pocket doors lead into the Family Room which has a fireplace

■ A tray ceiling, whirlpool tub, separate shower, double vanity and a walk-in closet enhance the Master Suite

■ Three additional bedrooms share a full bath

■ No materials list available

FIRST FLOOR — 1,194 SQ. FT.
SECOND FLOOR — 1,065 SQ. FT.
GARAGE — 672 SQ. FT.

Refer to **Pricing Schedule D** on the order form for pricing information

A Warm Welcome

■ This plan features:

— Three bedrooms

— Two full and one half baths

■ A large, island Kitchen with a built-in pantry, built-in desk and a double sink

■ A vaulted ceiling in the Sun Room

■ A tray ceiling in the formal Dining Room with easy access to the Kitchen

■ A fireplace and a built-in wetbar in the informal Family Room

■ A vaulted ceiling in the Master Suite which is equipped with his-n-her walk-in closets and a private full Bath

■ A barrel vaulted ceiling in the front bedroom

■ A convenient second floor Laundry Room

FIRST FLOOR — 1,336 SQ. FT.
SECOND FLOOR — 1,015 SQ. FT.
BASEMENT — 1,336 SQ. FT.
GARAGE — 496 SQ. FT.

TOTAL LIVING AREA:
2,351 SQ. FT.

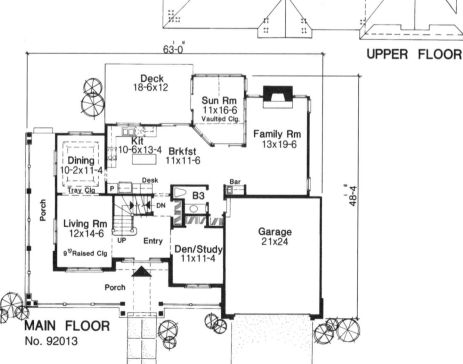

UPPER FLOOR

MAIN FLOOR
No. 92013

To order your Blueprints, call 1-800-235-5700

An
EXCLUSIVE DESIGN
By Karl Kreeger

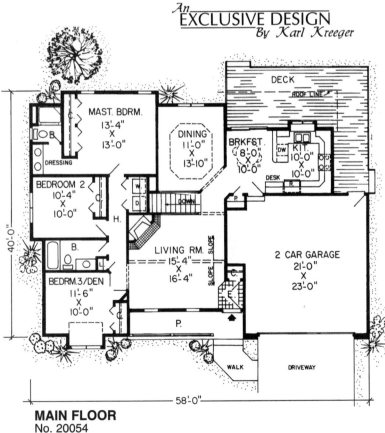

MAIN FLOOR
No. 20054

Striking Entryway

■ This plan features:

— Two bedrooms, with optional third bedroom/den

— Two full baths

■ A cathedral ceiling gracing the Living Room

■ A large Master Bedroom with an ample closet and a full Master Bath

■ A Dining Room with an attractive decorative ceiling

■ A modern Kitchen flowing into the Breakfast area

■ A conveniently located laundry area

MAIN AREA — 1,461 SQ. FT.
BASEMENT — 1,458 SQ. FT.
GARAGE — 528 SQ. FT.

TOTAL LIVING AREA:
1,461 SQ. FT.

Refer to **Pricing Schedule C** on
the order form for pricing information

Two-Sink Baths Ease Rush

■ This plan features:

— Four bedrooms

— Two full and one half baths

■ A wood beam ceiling in the spacious Family Room

■ An efficient, island Kitchen with a sunny bay window dinette

■ A formal Living Room with a heat-circulating fireplace

■ A large Master Suite with a walk-in closet and a private Master Bath

■ Three additional bedrooms sharing a full hall bath

FIRST FLOOR — 983 SQ. FT.
SECOND FLOOR — 1,013 SQ. FT.
MUDROOM — 99 SQ. FT.
GARAGE — 481 SQ. FT.

TOTAL LIVING AREA:
2,095 SQ. FT.

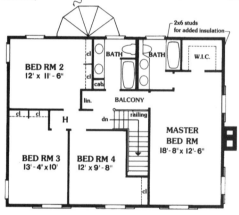

SECOND FLOOR PLAN

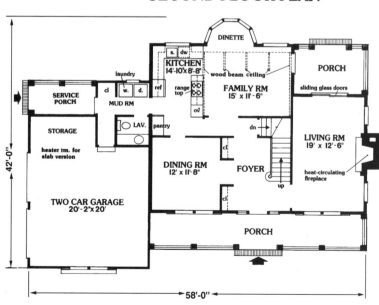

FIRST FLOOR PLAN
No. 90622

To order your Blueprints, call 1-800-235-5700

© 1993 Donald A. Gardner Architects, Inc.

B. NATHAN

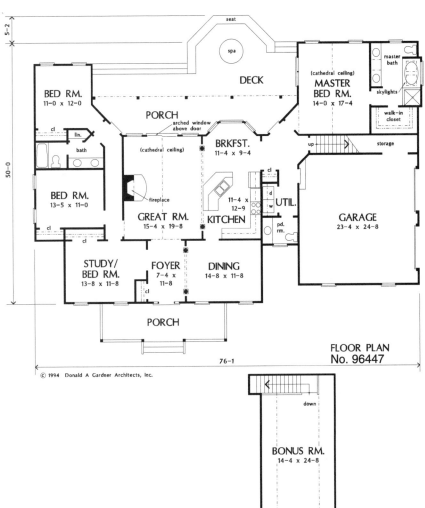

FLOOR PLAN
No. 96447

© 1994 Donald A Gardner Architects, Inc.

76-1

BONUS RM.
14-4 x 24-8

Refined Country Style

■ This plan features:

— Four bedrooms

— Two full and one half baths

■ Arched windows and interior columns add refined style

■ Generous Great Room with soaring cathedral ceiling

■ Well planned Kitchen with adjoining nook and angles counter

■ Privately situated Master Bedroom with skylit bath and walk-in closet

■ Large secondary bedrooms share a full bath

MAIN FLOOR — 2,207 SQ. FT.
BONUS AREA — 435 SQ. FT.
GARAGE — 634 SQ. FT.

TOTAL LIVING AREA:
2,207 SQ. FT.

Refer to **Pricing Schedule B** on the order form for pricing information

Outdoor-Lovers' Delight

■ This plan features:

— Three bedrooms

— Two full baths

■ A roomy Kitchen and Dining Room

■ A massive Living Room with a fireplace and access to the wrap-around porch via double French doors

■ An elegant Master Suite and two additional spacious bedrooms closely located to the laundry area

MAIN FLOOR — 1,540 SQ. FT.
PORCHES — 530 SQ. FT.

TOTAL LIVING AREA:
1,540 SQ. FT.

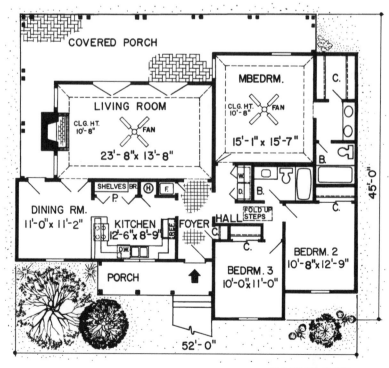

MAIN FLOOR
No. 10748

To order your Blueprints, call 1-800-235-5700

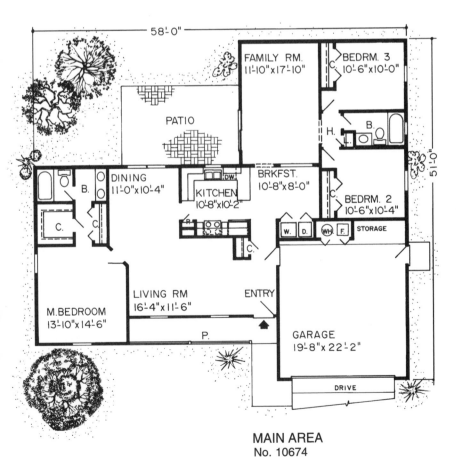

FAMILY RM.
11'-10"x17'-10"

BEDRM. 3
10'-6"x10'-0"

PATIO

B.

H.

DINING
11'-0"x10'-4"

B.

BRKFST.
10'-8"x8'-0"

KITCHEN
10'-8"x10'-2"

C.

C.

BEDRM. 2
10'-6"x10'-4"

W. D.

WH F.

STORAGE

C.

LIVING RM
16'-4"x11'-6"

ENTRY

M.BEDROOM
13'-10"x14'-6"

P.

GARAGE
19'-8"x22'-2"

DRIVE

58'-0"

51'-0"

MAIN AREA
No. 10674

Carefree Convenience

■ This plan features:

— Three bedrooms

— Two full baths

■ A galley Kitchen, centrally
located between the Dining,
Breakfast and Living Room areas

■ A huge Family Room which exits
onto the patio

■ A Master Suite with double
closets and vanities with two
additional bedrooms share a
full-half bath

MAIN AREA — 1,600 SQ. FT.
GARAGE — 465 SQ. FT.

TOTAL LIVING AREA:
1,600 SQ. FT.

To order your Blueprints, call 1-800-235-5700

Refer to **Pricing Schedule C** on the order form for pricing information

Reminiscent of the Deep South

■ This plan features:

— Three bedrooms

— Two full and one half baths

■ Victorian Porch leads into Foyer and two-story Great Room with a focal point fireplace

■ Efficient, U-shaped Kitchen opens to Great Room and Dining Room

■ Private Master Suite with triple windows, walk-in closet and luxurious bath with a double vanity and garden window tub

■ Two second floor bedrooms with large closets, share a full bath and Loft

FIRST FLOOR — 1,350 SQ. FT.
SECOND FLOOR — 589 SQ. FT.

TOTAL LIVING AREA:
1,939 SQ. FT.

An
EXCLUSIVE DESIGN
By United Design Associates

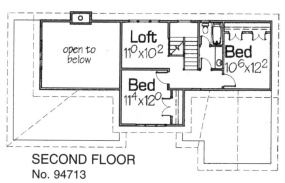

SECOND FLOOR
No. 94713

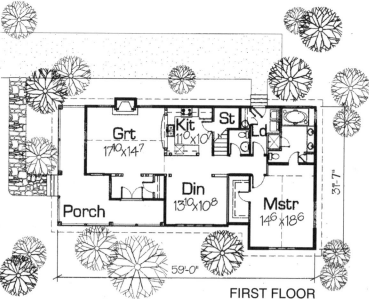

FIRST FLOOR

To order your Blueprints, call 1-800-235-5700

FIRST FLOOR
No. 93333

FLOOR, ABOVE

BOOKS

SEAT

DEN
14 x 13

SUN RM
13 x 12

DECK

KITCHEN
13 x 14

DINETTE
12 x 11-6

CIRCLE-HEAD
WINDOW

TRAY CLG
FAMILY
16 x 22

RAILING

DN

94-6

60-2

LIVING
14 x 19

OPEN
ABOVE

UP
DN

FOYER

REF
OV

PAN.

DINING
14 x 14
STEPPED CLG

ENTRY

P

GARAGE
24 x 34(+)

8-FT DOORS

PORCH

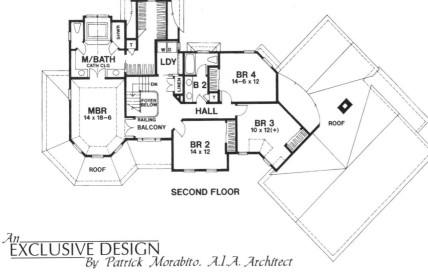

SECOND FLOOR

M/BATH
CATH CLG

LDY

W D

B 2

BR 4
14-6 x 12

DN

FOYER
BELOW

LINEN

T

MBR
14 x 18-6

RAILING
BALCONY

HALL

BR 3
10 x 12(+)

ROOF

BR 2
14 x 12

ROOF

A Whisper of Victorian Styling

■ This plan features:

— Four bedrooms

— Two full and one half baths

■ Formal Living Room features wrap-around windows and direct access to the front Porch

■ An elegant, formal Dining Room accented by a stepped ceiling

■ A bright, all-purpose Sun Room adjoins an expansive Deck

■ Family Room, with a tray ceiling topping a circle head window and a massive, hearth fireplace

■ Private Master Suite has a decorative ceiling and a luxurious Bath with a raised, atrium tub

■ No materials list available

FIRST FLOOR — 1,743 SQ. FT.
SECOND FLOOR — 1,455 SQ. FT.

TOTAL LIVING AREA:
3,198 SQ. FT.

Refer to **Pricing Schedule E** on the order form for pricing informatior

Classic Victorian

■ This plan features:

— Four bedrooms

— Three full and one half baths

■ Large open areas that are bright and free flowing

■ Great Room accented by a fireplace and large front window

■ Sun Room off of the Great Room viewing the porch

■ Dining Room in close proximity to the Kitchen

■ Efficient Kitchen flows into informal Breakfast Nook

■ Private first floor Master Suite highlighted by a plush Master Bath

■ Three bedrooms on the second floor, two with walk-in closets and one with a private bath

FIRST FLOOR—1,868 SQ. FT.
SECOND FLOOR—964 SQ. FT.
GARAGE—460 SQ. FT.

TOTAL LIVING AREA:
2,832 SQ. FT.

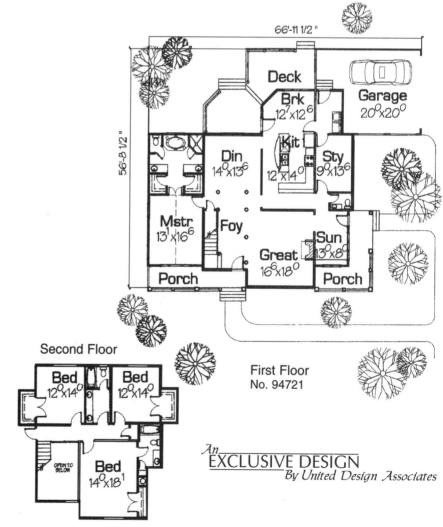

First Floor
No. 94721

Second Floor

An EXCLUSIVE DESIGN
By United Design Associates

Refer to **Pricing Schedule C** on the order form for pricing information

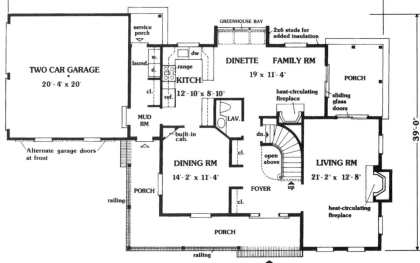

FIRST FLOOR PLAN
No. 90687

TWO CAR GARAGE
20'- 4' x 20'

service porch

laund.

w.
d.

KITCH.
12'-10' x 8'-10'

range

dw

ref.

cl.

MUD RM

GREENHOUSE BAY

2x6 studs for added insulation

DINETTE

FAMILY RM
19' x 11'- 4'

PORCH

heat-circulating fireplace

sliding glass doors

LAV.

built-in cab.

dn.

open above

up

DINING RM
14'- 2' x 11'- 4'

cl.

FOYER

cl.

LIVING RM
21'- 2' x 12'- 8'

heat-circulating fireplace

Alternate garage doors at front

PORCH

railing

PORCH

railing

64'-0"

39'-0"

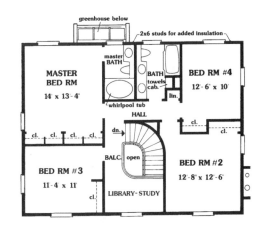

SECOND FLOOR PLAN

greenhouse below

2x6 studs for added insulation

MASTER BED RM
14' x 13'- 4'

master BATH

whirlpool tub

BATH

towels cab.

lin.

BED RM #4
12'- 6' x 10'

cl. cl. cl. cl.

HALL

dn.

open

cl.

BED RM #3
11'- 4' x 11'

BALC.

cl.

LIBRARY - STUDY

BED RM #2
12'- 8' x 12'- 6'

Country Comforts

■ This plan features:

— Four bedrooms

— Two and one half baths

■ A covered porch, window boxes, and two chimneys

■ Cozy Living and Dining Rooms

■ Cabinets and a greenhouse bay separate the Kitchen, dinette, and Family Room overlooking the backyard

■ A covered porch just off the fireplaced Family Room

FIRST FLOOR — 1,065 SQ. FT.
SECOND FLOOR — 1,007 SQ. FT.
LAUNDRY/MUDROOM — 88 SQ. FT.
GARAGE — 428 SQ. FT.

TOTAL LIVING AREA:
2,160 SQ. FT.

Refer to **Pricing Schedule C** on the order form for pricing information

© 1990 Donald A. Gardner, Architects, Inc

Flexibility to Expand

■ This plan features:

— Three bedrooms

— Two full and one half baths

■ Two-story Foyer contains palladian window in a clerestory dormer

■ Efficient Kitchen opens to Breakfast area and Deck for outdoor dining

■ Columns separating the Great Room and the Dining Room which have nine foot ceilings

■ First level Master Bedroom Suite features a skylight above the whirlpool tub in the bath

■ Please specify basement or crawl space foundation when ordering this plan

FIRST FLOOR — 1,289 SQ. FT.
SECOND FLOOR — 542 SQ. FT.
BONUS ROOM — 393 SQ. FT.
GARAGE & STORAGE — 521 SQ. FT.

TOTAL LIVING AREA:
1,831 SQ. FT.

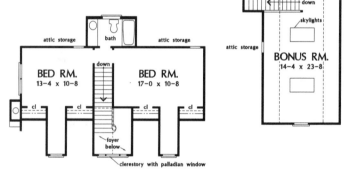

SECOND FLOOR PLAN

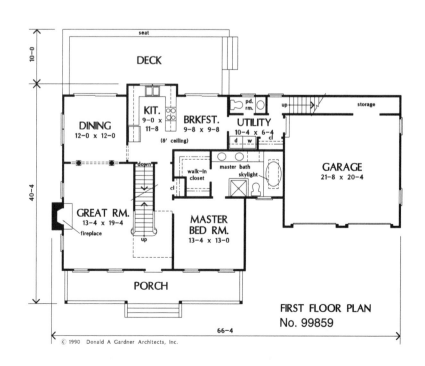

FIRST FLOOR PLAN
No. 99859

© 1990 Donald A Gardner Architects, Inc.

To order your Blueprints, call 1-800-235-5700

An EXCLUSIVE DESIGN
By Jannis Vann & Associates, Inc.

A Touch of Victorian Styling

■ This plan features:

— Three bedrooms

— Two full and one half baths

■ A covered porch and a pointed roof on the sitting alcove of the Master Suite giving this home a Victorian look

■ A formal Living Room directly across from the Dining Room for ease in entertaining

■ An efficient Kitchen with a bright bayed Breakfast area

■ The Family Room has a cozy fireplace nestled in a corner

■ A large Master Suite with a cozy sitting alcove and double vanity bath

■ Two additional bedrooms serviced by a full hall bath

FIRST FLOOR — 887 SQ. FT.
SECOND FLOOR — 877 SQ. FT.
BASEMENT — 859 SQ. FT.
GARAGE — 484 SQ. FT.

SECOND FLOOR

BDRM. 3
11'·6"x10'·2"

H. BATH

M. BATH

BDRM. 2
11'·6"x12'·2"

M. BDRM.
11'·6"x18'·6"

SITTING ALCOVE

■ No materials list is available for this plan

TOTAL LIVING AREA:
1,764 SQ. FT.

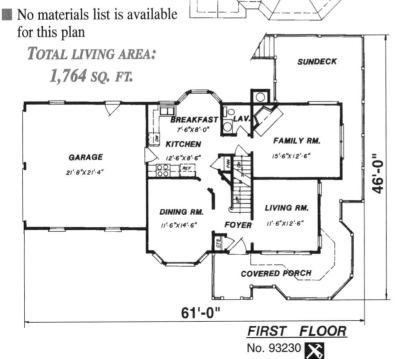

SUNDECK

BREAKFAST
7'·6"x8'·0"

LAV.

KITCHEN
12'·6"x8'·6"

FAMILY RM.
15'·6"x12'·6"

GARAGE
21'·8"x21'·4"

REF.

DINING RM.
11'·6"x14'·6"

LIVING RM.
11'·6"x12'·6"

FOYER

COVERED PORCH

46'-0"

61'-0"

FIRST FLOOR
No. 93230

Refer to **Pricing Schedule D** on the order form for pricing information

Gingerbread Gem

■ This plan features:

— Three bedrooms

— Two and one half baths

■ Warm weather living space afforded by a wrap-around covered porch, second floor balcony, and a huge rear deck

■ Large windows adding a sunny glow to every room

■ The family areas flowing together for a wide open atmosphere that is warmed by the Family Room fireplace

■ A large Master Suite with an abundance of closet space and an expansive bath area

FIRST FLOOR — 1,311 SQ. FT.
SECOND FLOOR — 968 SQ. FT.
BASEMENT — 968 SQ. FT.

TOTAL LIVING AREA:
2,279 SQ. FT.

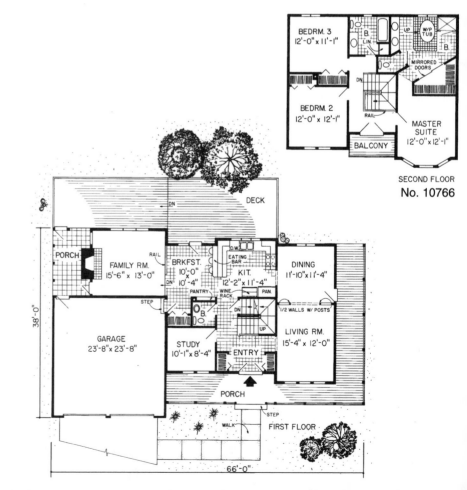

SECOND FLOOR
No. 10766

To order your Blueprints, call 1-800-235-5700

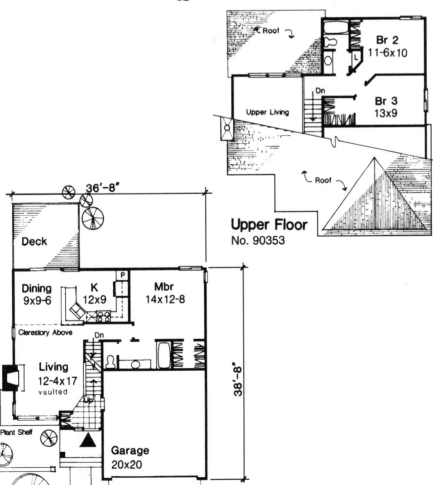

Upper Floor
No. 90353

Br 2
11-6x10

Br 3
13x9

Upper Living

Roof

36'-8"

Deck

Dining
9x9-6

K
12x9

Mbr
14x12-8

Clerestory Above

Living
12-4x17
vaulted

Plant Shelf

Garage
20x20

38'-8"

Main Floor

Living Room Features Vaulted Ceiling

- This plan features:
- — Three bedrooms
- — Two full baths
- A vaulted ceiling in the Living Room and the Dining Room, with a clerestory above
- A Master Bedroom with a walk-in closet and private full bath
- An efficient Kitchen, with a corner double sink and peninsula counter
- A Dining Room with sliding doors to the Deck
- A Living Room with a fireplace that adds warmth to open areas
- Two additional bedrooms that share a full hall bath

FIRST FLOOR — 846 SQ. FT.
SECOND FLOOR — 400 SQ. FT.

TOTAL LIVING AREA:
1,246 SQ. FT.

Refer to **Pricing Schedule C** on the order form for pricing information

Streetside Appeal

■ This plan features:

— Three bedrooms

— Two full and one half baths

■ An elegant Living and Dining Room combination that is divided by columns

■ A Family/Hearth Room with a two-way fireplace to the Breakfast room

■ A well-appointed Kitchen with built-in pantry, peninsula counter and double corner sink

■ A Master Suite with decorative ceiling, walk-in closet and private bath

■ Two additional bedrooms that share a full hall bath

FIRST FLOOR — 1,590 SQ. FT.
SECOND FLOOR — 567 SQ. FT.
BASEMENT — 1,576 SQ. FT.
GARAGE — 456 SQ. FT.

TOTAL LIVING AREA:
2,157 SQ. FT.

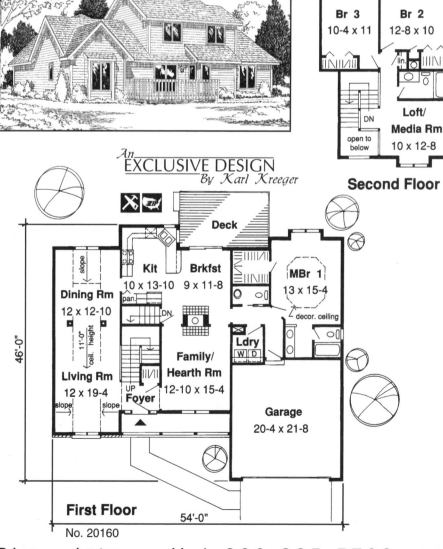

An
EXCLUSIVE DESIGN
By Karl Kreeger

Second Floor

Br 3
10-4 x 11

Br 2
12-8 x 10

Loft/
Media Rm
10 x 12-8

DN

open to below

Deck

Kit
10 x 13-10

Brkfst
9 x 11-8

MBr 1
13 x 15-4

decor. ceiling

Dining Rm
12 x 12-10

pan.

DN

Ldry
W D

Living Rm
12 x 19-4

slope

UP
Foyer

Family/
Hearth Rm
12-10 x 15-4

Garage
20-4 x 21-8

11'-0" ceil. height

46'-0"

slope

First Floor
54'-0"

No. 20160

FIRST FLOOR — 2,804 SQ. FT.
SECOND FLOOR — 979 SQ. FT.
BASEMENT — 2,804 SQ. FT.
GARAGE — 802 SQ. FT.

TOTAL LIVING AREA:
3,783 SQ. FT.

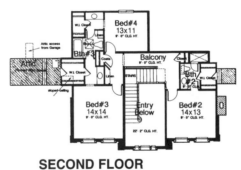

SECOND FLOOR

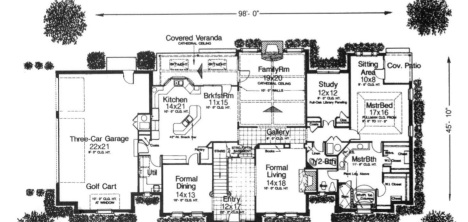

FIRST FLOOR
No. 92237

Opulent Luxury

◼ This plan features:

— Four bedrooms

— Three full and one half baths

◼ Magnificent columns frame elegant two-story Entry with a graceful banister staircase

◼ A stone hearth fireplace and built-in book shelves enhance the Living Room

◼ Comfortable Family Room with a huge fireplace, cathedral ceiling and access to Covered Veranda

◼ Spacious Kitchen with cooktop island/snackbar, built-in pantry and Breakfast Room

◼ Lavish Master Bedroom wing with a pullman ceiling, sitting area, private Covered Patio and a huge bath with two walk-in closets and a whirlpool tub

◼ Three additional bedrooms on second floor with walk-in closets and private access to a full bath

◼ No materials list available

Country Victorian

■ This plan features:

— Three bedrooms

— Two full and one half baths

■ Covered Entry leads into an open Foyer with a landing staircase and balcony

■ The Living and Dining rooms connect for entertaining ease

■ Efficient Kitchen with a work island, built-in pantry and Dining area

■ Expansive Family Room has a cathedral ceiling and a fireplace

■ Master Bedroom is enhanced by a walk-in closet and whirlpool tub

■ Two additional bedrooms share a full bath

■ No materials list available

FIRST FLOOR — 1,228 SQ. FT.
SECOND FLOOR — 952 SQ. FT.
BASEMENT — 1,228 SQ. FT.
GARAGE — 479 SQ. FT.

TOTAL LIVING AREA:
2,180 SQ. FT.

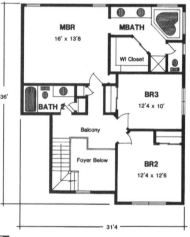

SECOND FLOOR

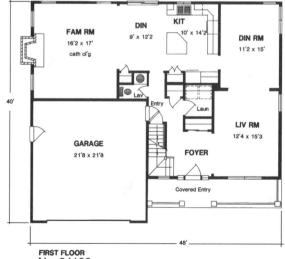

FIRST FLOOR
No. 94100

To order your Blueprints, call 1-800-235-5700

TOTAL LIVING AREA:
1,700 sq. ft.

Main Floor
No. 24250

An
EXCLUSIVE DESIGN
By Energetic Enterprises

Clever Design Packs in Plenty of Living Space

■ This plan features:

— Three bedrooms

— Two full baths

■ Custom, volume ceilings

■ A sunken Living Room that includes a vaulted ceiling and a fireplace with oversized windows framing it

■ A center island and an eating nook in the Kitchen that has more than ample counter space

■ A formal Dining Room that adjoins the Kitchen, allowing for easy entertaining

■ A spacious Master Suite including a vaulted ceiling and lavish bath

■ Secondary bedrooms with custom ceiling treatments and use of full hall bath

MAIN AREA — 1,700 sq. ft.

Floor Plan Labels

55'-4"

Optional Patio

Nook
15-6 x 8
8'-9" clg.
plant shelf

Living Rm
vault clg.
slope
slope
8' clg.

Kit.
15-6 x 10-8
14-8 x 18-8

win. seat

MBr
13-6 x 16
vault clg.

glass block

Dining Rm
8'-9" clg.
14-2 x 10-4

10' clg.
1/2 wall
railing
lin.

Foyer

Br. 2
11-10 x 10-8

Br. 3
11-10 x 10-8

Porch

Garage
20 x 21

53'-3 1/2"

Country Ranch

■ This plan features:

— Three bedrooms

— Two full baths

■ A railed and covered wrap-around porch, adding charm to this country-styled home

■ A high vaulted ceiling in the Living Room

■ A smaller Kitchen with ample cupboard and counter space, that is augmented by a large pantry

■ An informal Family Room with access to the wood deck

■ A private Master Suite with a spa tub and a walk-in closet

■ Two family bedrooms that share a full hall bath

■ A shop and storage area in the two-car garage

MAIN AREA — 1,485 SQ. FT.
GARAGE — 701 SQ. FT.

TOTAL LIVING AREA: *1,485* SQ. FT.

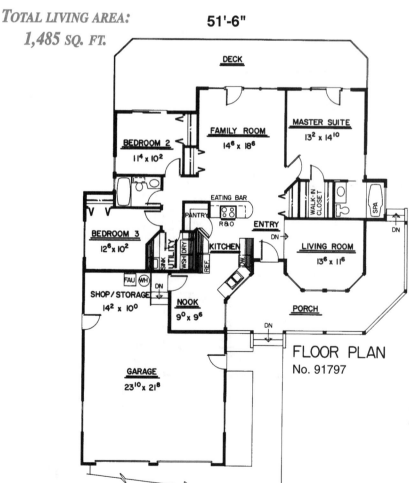

FLOOR PLAN
No. 91797

PLAN NO. 94900

SECOND FLOOR

Br. 2
11⁰ x 11⁴

Br. 4
11⁰ x 10⁰

DESK

OPEN
TO
BELOW

DN

Br. 3
11³ x 11³
10'-0"
CEILING

UNFINISHED
STORAGE
14⁶ x 12⁶

© design basics, inc.

FIRST FLOOR
No. 94900

TRANSOMS

TRANS.

TRANS.

LIN.

WHIRLPOOL

Grt. rm.
15³ x 19⁹
12'-10" CEILING

Bfst.
12⁶ x 13⁷
SNACK BAR

Kit.
10⁰ x 11³
CEILING

SHELVES

DESK

R.

P.

UP

DN

D. W.

Mbr.
13⁰ x 16³
11'-6" CLG.

E.

Din.
12³ x 12⁸

HUTCH

Gar.
20⁸ x 23⁰

COVERED
PORCH

TRANSOMS

BENCH

47'-4"

52'-0"

Quaint Front Porch and Lovely Details

■ This plan features:

— Four bedrooms

— Two full and one half baths

■ A Covered Porch and Victorian touches create unique elevation

■ A one and a half story entry hall leads into formal Dining Room

■ A volume ceiling above abundant windows and a see-through fireplace highlight the Great Room

■ Kitchen/Breakfast area shares the fireplace and has a snack bar, desk, walk-in pantry and abundant counter space

■ Laundry area provides access to Garage and side yard also

■ Secluded Master Suite crowned by a vaulted ceiling and a luxurious bath

■ Three additional bedrooms on the second floor share a full bath

FIRST FLOOR — 1,421 SQ. FT.
SECOND FLOOR — 578 SQ. FT.
BASEMENT — 1,421 SQ. FT.
GARAGE — 480 SQ. FT.

TOTAL LIVING AREA :
1,999 SQ. FT.

Refer to **Pricing Schedule C** on the order form for pricing information

© 1994 Donald A. Gardner Architects, Inc

Mixture of Traditional and Country Charm

■ This plan features:

— Three bedrooms

— Two full and one half baths

■ Stairs to the skylit bonus room located near the Kitchen and Master Suite

■ Master Suite crowned in cathedral ceilings has a skylit bath containing a whirlpool tub and dual vanity

■ Great Room, topped by a cathedral ceiling and highlighted by a fireplace, is adjacent to the country Kitchen

■ Two additional bedrooms share a hall bath

MAIN FLOOR — 1,954 SQ. FT.
GARAGE — 649 SQ. FT.

TOTAL LIVING AREA:
1,954 SQ. FT.

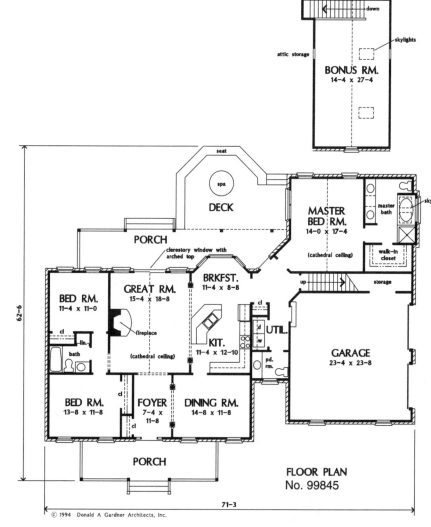

BONUS RM.
14-4 x 27-4

attic storage

skylights

down

seat

spa

DECK

PORCH

clerestory window with arched top

GREAT RM.
15-4 x 18-8

fireplace

(cathedral ceiling)

BRKFST.
11-4 x 8-8

MASTER BED RM.
14-0 x 17-4

(cathedral ceiling)

master bath

skylights

walk-in closet

storage

up

KIT.
11-4 x 12-10

UTIL.

pd. rm.

GARAGE
23-4 x 23-8

BED RM.
11-4 x 11-0

lin.

bath

cl

BED RM.
13-8 x 11-8

FOYER
7-4 x 11-8

DINING RM.
14-8 x 11-8

PORCH

62-6

71-3

FLOOR PLAN
No. 99845

© 1994 Donald A Gardner Architects, Inc.

To order your Blueprints, call 1-800-235-5700

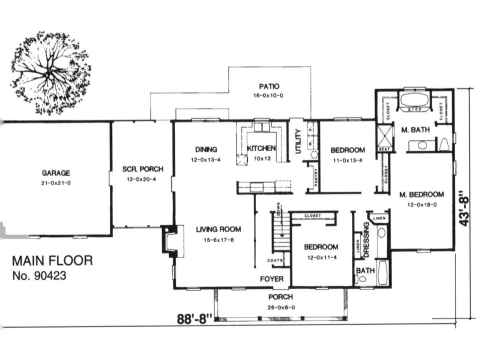

MAIN FLOOR
No. 90423

PATIO
16-0x10-0

DINING
12-0x13-4

KITCHEN
10x13

UTILITY

BEDROOM
11-0x13-4

M. BATH

CLOSET

STEP

CLOSET

SEAT

GARAGE
21-0x21-0

SCR. PORCH
12-0x20-4

PANTRY

M. BEDROOM
12-0x18-0

LIVING ROOM
15-6x17-8

DOWN

CLOSET

COATS

BEDROOM
12-0x11-4

DRESSING

LINEN

LINEN

BATH

FOYER

PORCH
26-0x6-0

88'-8"

43'-8"

Expansive, Not Expensive

■ This plan features:

— Three bedrooms

— Two full baths

■ A Master Suite with his-n-her closets and a private Master Bath

■ Two additional bedrooms that share a full hall closet

■ A pleasant Dining Room that overlooks a rear garden

■ A well-equipped Kitchen with a built-in planning corner and eat-in space

■ This plan is available with a basement, slab or crawl space foundation — please specify

MAIN FLOOR — 1,773 SQ. FT.

TOTAL LIVING AREA:
1,773 SQ. FT.

109

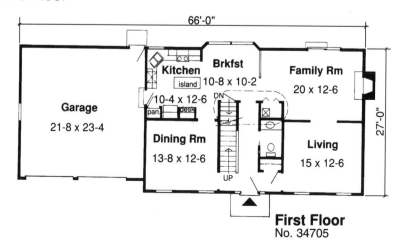

Refer to **Pricing Schedule D** on the order form for pricing information

Colonial Home with All the Traditional Comforts

■ This plan features:

— Four bedrooms

— Two and one half baths

■ A formal Living Room and Dining Room flanking a spacious entry

■ Family areas flowing together into an open space at the rear of the home

■ An island Kitchen with a built-in pantry centrally located for easy service to the Dining Room and Breakfast area

■ A Master Suite with large closets and a double vanity in the bath

FIRST FLOOR — 1,090 SQ. FT.
SECOND FLOOR — 1,134 SQ. FT.
BASEMENT — 1,090 SQ. FT.
GARAGE — 576 SQ. FT.

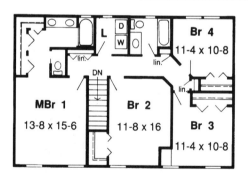

Second Floor

Br 4
11-4 x 10-8

MBr 1
13-8 x 15-6

Br 2
11-8 x 16

Br 3
11-4 x 10-8

Slab/Crawlspace Option

TOTAL LIVING AREA:
2,224 SQ. FT.

66'-0"

27'-0"

Garage
21-8 x 23-4

Kitchen
10-4 x 12-6

island

pan.   desk

Brkfst
10-8 x 10-2

Family Rm
20 x 12-6

Dining Rm
13-8 x 12-6

Living
15 x 12-6

DN

UP

First Floor
No. 34705

To order your Blueprints, call 1-800-235-5700

© 1997 Donald A. Gardner Architects, Inc.

B. NATHAN

MAIN FLOOR — 2,057 SQ. FT.
GARAGE & STORAGE — 622 SQ. FT.
BONUS ROOM — 444 SQ. FT.

TOTAL LIVING AREA:
2,057 SQ. FT.

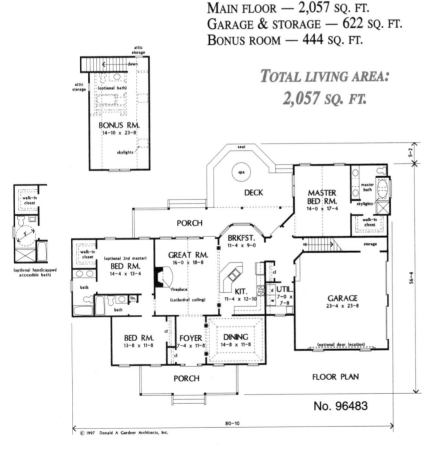

No. 96483

© 1997 Donald A Gardner Architects, Inc.

Grace and Style

■ This plan features:

— Three bedrooms

— Three full baths

■ Foyer accented by columns gives entry into the formal Dining Room

■ Angled island Kitchen is open to the Breakfast Bay

■ Great Room topped by a cathedral ceiling and enhanced by a fireplace and access to the rear porch

■ Secluded Master Suite with a skylit bath

■ Two secondary bedrooms, one with a private bath, an alternate bath design creates a wheel chair accessibility option for the disabled

■ Bonus room may create a terrific fourth bedroom and bath

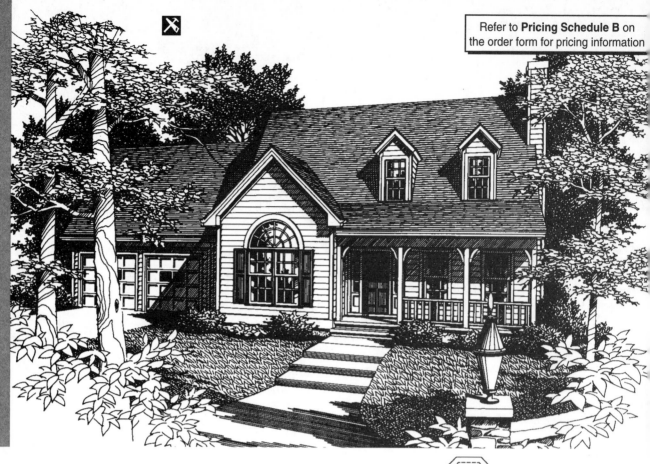

Refer to **Pricing Schedule B** on the order form for pricing information

Charming, Compact and Convenient

■ This plan features:

— Three bedrooms

— Two full and one half bath

■ Double dormer, arched window and Covered Porch add light and space

■ Open Foyer graced by banister staircase and balcony

■ Spacious Activity Room with a pre-fab fireplace opens to formal Dining Room

■ Country-size Kitchen/Breakfast area with island counter and access to Sun Deck and Laundry/Garage entry

■ First floor bedroom highlighted by lovely arched window below a tray ceiling and a pampering bath

■ Two upstairs bedrooms share a twin vanity bath

■ When ordering this plan —please specify a basement or crawlspace foundation

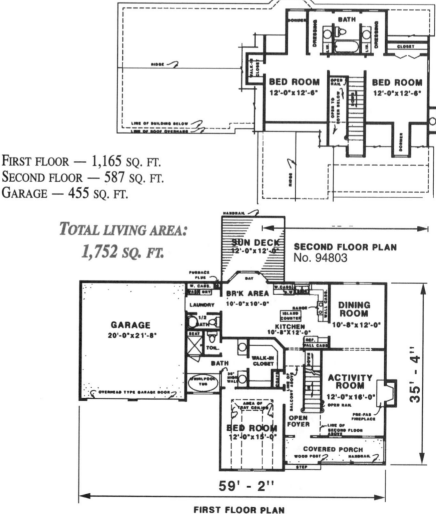

FIRST FLOOR — 1,165 SQ. FT.
SECOND FLOOR — 587 SQ. FT.
GARAGE — 455 SQ. FT.

TOTAL LIVING AREA:
1,752 SQ. FT.

To order your Blueprints, call 1-800-235-5700

Refer to **Pricing Schedule D** on the order form for pricing information

SECOND FLOOR — 687 SQ. FT.

TOTAL LIVING AREA:
2,562 SQ. FT.

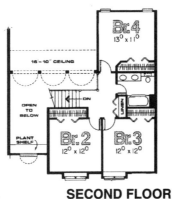

SECOND FLOOR

© design basics, inc.

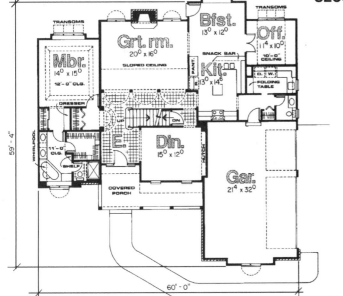

FIRST FLOOR
No. 99432

Arched Openings in Great Room

■ This plan features:

—Four bedrooms

—Two full and one half baths

■ Dazzling fifteen foot arched openings accent the Great Room's entry

■ Fireplace with windows to either side enhances the Great Room

■ Formal dining room easily accessible from the large island Kitchen

■ Elegant French doors in the Breakfast Room open to the versatile office topped by a ten foot ceiling

■ Luxurious Master Suite with a private entrance, volume ceiling, a built-in dresser, two closets and a beautiful corner whirlpool

■ Three additional bedrooms, on the second floor, share a bath with a double vanity

FIRST FLOOR — 1,879 SQ. FT.

Refer to **Pricing Schedule B** on the order form for pricing information

Economy at It's Best

■ This plan features:

— Three bedrooms

— Two full and one three quarter bath

■ Attractive porch adds to the curb appeal of this economical to build home

■ A vaulted ceiling topping the Entry, Living and Dining Rooms

■ A lovely bay window, adding sophistication to the Living Room

■ A Master Suite with a walk-in closet, and a private compartmented bath with an over-sized shower

■ Two additional bedrooms share a full hall bath topped by a skylight

■ A walk-in pantry adds to the storage space of the cooktop island Kitchen, which is equipped with a double sink with a window above

■ Garage offers direct entrance to the house

MAIN AREA — 1,717 SQ. FT.
GARAGE — 782 SQ. FT.
WIDTH — 80'– 0"
DEPTH — 42'– 0"

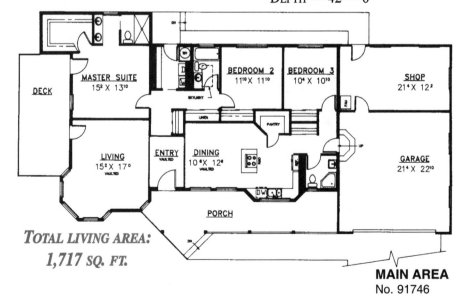

TOTAL LIVING AREA:
1,717 SQ. FT.

MAIN AREA
No. 91746

FIRST FLOOR
No. 94905

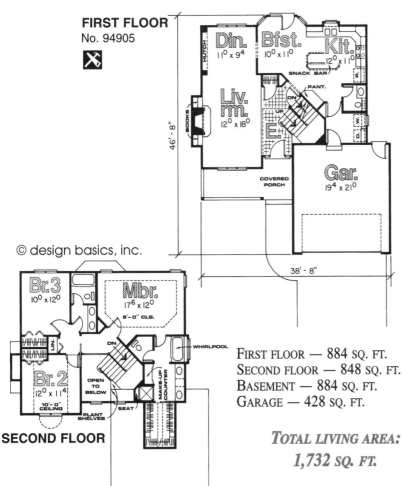

© design basics, inc.

FIRST FLOOR — 884 SQ. FT.
SECOND FLOOR — 848 SQ. FT.
BASEMENT — 884 SQ. FT.
GARAGE — 428 SQ. FT.

TOTAL LIVING AREA:
1,732 SQ. FT.

Angled Staircase Accents Entry

■ This plan features:

— Three bedrooms

— Two full and one half baths

■ Two-story Entry with sidelights leads to open Living Room/Dining area

■ Built-ins and decorative windows highlight Living and Dining Rooms

■ Bayed Breakfast area adds to open Kitchen with a large pantry, island counter/snack bar and convenient access to laundry room

■ A great Master Bedroom suite has a whirlpool bath with double vanity and a generous walk-in closet

■ Bedroom two contains beautiful arched window below a volume ceiling

■ Secondary bedrooms share a double vanity bath

©1995 Donald A. Gardner Architects, Inc.

Refer to **Pricing Schedule D** on the order form for pricing information

Country Charm and Modern Convenience

■ This plan features:

— Three bedrooms

— Two full and one half baths

■ Great Room crowned by a cathedral ceiling and accented by a cozy fireplace with built-ins

■ Centrally located Kitchen with nearby pantry serving Breakfast area and Dining Room with ease

■ Master Suite elegantly appointed by a walk-in closet and a lavish bath

■ A Sitting Room with bay window off the Master Suite

■ Two secondary bedrooms sharing a full bath

FIRST FLOOR — 1,778 SQ. FT.
SECOND FLOOR — 592 SQ. FT.
GARAGE & STORAGE — 622 SQ. FT.
BONUS ROOM — 404 SQ. FT.

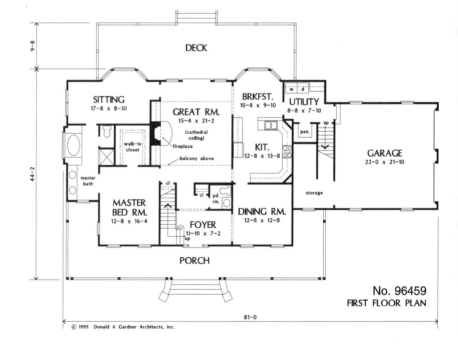

No. 96459
FIRST FLOOR PLAN

© 1995 Donald A Gardner Architects, Inc.

SECOND FLOOR PLAN

TOTAL LIVING AREA:
2,370 SQ. FT.

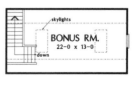

To order your Blueprints, call 1-800-235-5700

TOTAL LIVING AREA:
1,7382 SQ. FT.

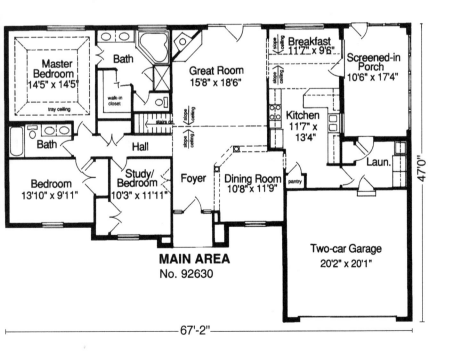

MAIN AREA
No. 92630

67'-2"

Charming Brick Ranch

■ This plan features:

— Three bedrooms

— Two full baths

■ Sheltered entrance leads into open Foyer and Dining Room defined by columns

■ Vaulted ceiling spans Foyer, Dining Room, and Great Room with corner fireplace and atrium door to rear year

■ Central Kitchen with separate Laundry and pantry easily serves Dining Room, Breakfast area and Screened Porch

■ Luxurious Master bedroom offers tray ceiling and French doors to double vanity, walk-in closet and whirlpool tub

■ Two additional bedrooms, one easily converted to a Study, share a full bath

■ No materials list available for this plan

MAIN FLOOR — 1,782 SQ. FT.
GARAGE — 407 SQ. FT.
BASEMENT — 1,735 SQ. FT.

Refer to **Pricing Schedule F** on the order form for pricing information

Brick Grandeur

■ This plan features:

— Four bedrooms

— Three full and one half baths

■ Dramatic two-story glass Entry with a curved staircase

■ Both Living and Family rooms offer high ceilings, decorative windows and large fireplaces

■ Large, but efficient Kitchen with a cooktop serving island, walk-in pantry, bright Breakfast area and Patio access

■ Lavish Master Bedroom with a cathedral ceiling, two walk-in closets, and large bath

■ Two additional bedrooms with ample closets, share a double vanity bath

■ No materials list available

FIRST FLOOR — 2,506 SQ. FT.
SECOND FLOOR — 1,415 SQ. FT.
GARAGE — 660 SQ. FT.

FIRST FLOOR
No. 92248

TOTAL LIVING AREA:
3,921 SQ. FT.

SECOND FLOOR

Refer to **Pricing Schedule C** on the order form for pricing information

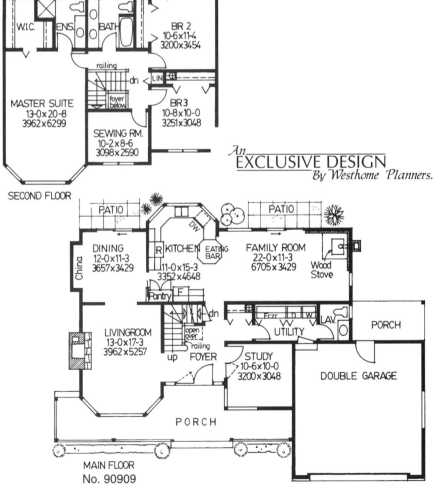

W.I.C. ENS. BATH

BR 2
10-6x11-4
3200x3454

railing
dn LIN
foyer
below

MASTER SUITE
13-0 x 20-8
3962x6299

BR 3
10-8 x 10-0
3251x3048

SEWING RM.
10-2 x 8-6
3098x2590

SECOND FLOOR

An
EXCLUSIVE DESIGN
By Westhome Planners.

PATIO   PATIO

DINING
12-0 x 11-3
3657x3429

KITCHEN
11-0 x 15-3
3352x4648

EATING
BAR

FAMILY ROOM
22-0 x 11-3
6705x3429

Wood
Stove

China

Pantry

LIVINGROOM
13-0 x 17-3
3962x5257

open
over
railing
up   FOYER

dn

Frzr   D   W
UTILITY
LAV.

PORCH

STUDY
10-6 x 10-0
3200x3048

DOUBLE GARAGE

PORCH

MAIN FLOOR
No. 90909

A Hint of Victorian Nostalgia

■ This plan features:

— Three bedrooms

— Two and one half baths

■ A classic center stairwell

■ A Kitchen with full bay window and built-in eating table

■ A spacious Master Suite including a large walk-in closet and full bath

FIRST FLOOR — 1,206 SQ. FT.
SECOND FLOOR — 969 SQ. FT.
GARAGE — 471 SQ. FT.
BASEMENT — 1,206 SQ. FT.
WIDTH — 61'
DEPTH — 44'

TOTAL LIVING AREA:
2,175 SQ. FT.

Refer to **Pricing Schedule C** on the order form for pricing information

Attractive Combination of Brick and Siding

■ This plan features:

— Three Bedrooms

— Two full Baths

■ A Great Room with sunny bayed area, fireplace and built-in entertainment center

■ A private Master Bedroom with luxurious Master Bath and walk-in closet

■ Dining Room has a Butler Pantry

■ Two additional bedrooms have use of hall full bath

MAIN LEVEL — 2,010 SQ. FT.
BASEMENT — 2,010 SQ. FT.

TOTAL LIVING AREA:
2,010 SQ. FT.

An EXCLUSIVE DESIGN
By Energetic Enterprises

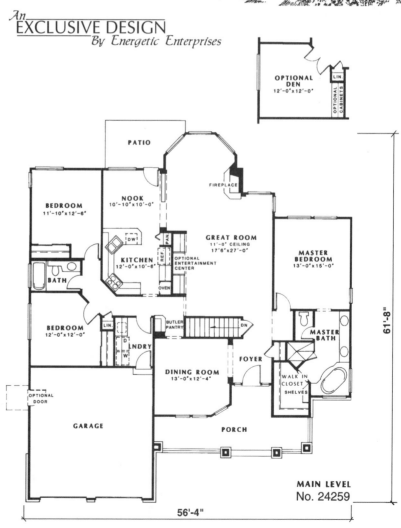

TOTAL LIVING AREA:
1,220 SQ. FT.

WIDTH 45'-0"
DEPTH 51'-6"

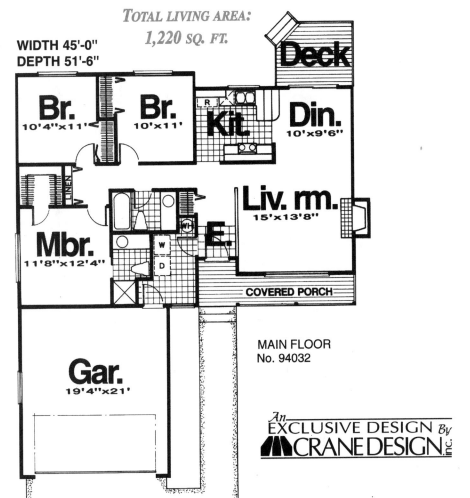

Deck

Br.
10'4"x11'

Br.
10'x11'

Kit.

Din.
10'x9'6"

LINEN

Liv. rm.
15'x13'8"

Mbr.
11'8"x12'4"

E.

WH

W

D

COVERED PORCH

Gar.
19'4"x21'

MAIN FLOOR
No. 94032

An
EXCLUSIVE DESIGN *By*
CRANE DESIGN inc.

Charming Country Porch Entry

◼ This plan features:

— Three bedrooms

— One full and one three-quarter baths

◼ Living/Dining room for ease in entertaining

◼ An inviting fireplace, double window and sliding glass door to the Deck

◼ Efficient Kitchen with a peninsula counter/snackbar, easily serves Dining area and Deck

◼ Roomy Master Bedroom features a walk-in closet and private bath

◼ Two additional bedrooms with ample closets, share a full bath

◼ Laundry facilities conveniently located near Garage and Entry

◼ No materials list available

MAIN FLOOR — 1,220 SQ. FT.
GARAGE — 440 SQ. FT.

Refer to **Pricing Schedule B** on the order form for pricing information

Compact Victorian
Ideal for Narrow Lot

■ This plan features:

— Three bedrooms

— Three full baths

■ A large, front Parlor with a raised hearth fireplace

■ A Dining Room with a sunny bay window

■ An efficient galley Kitchen serving the formal Dining Room and informal Breakfast Room

■ A beautiful Master Suite with two closets, an oversized tub and double vanity, plus a private sitting room with a bayed window and vaulted ceiling

■ Optional basement, crawl space or slab foundation — please specify when ordering

FIRST FLOOR — 954 SQ. FT.
SECOND FLOOR — 783 SQ. FT.

TOTAL LIVING AREA:
1,737 SQ. FT.

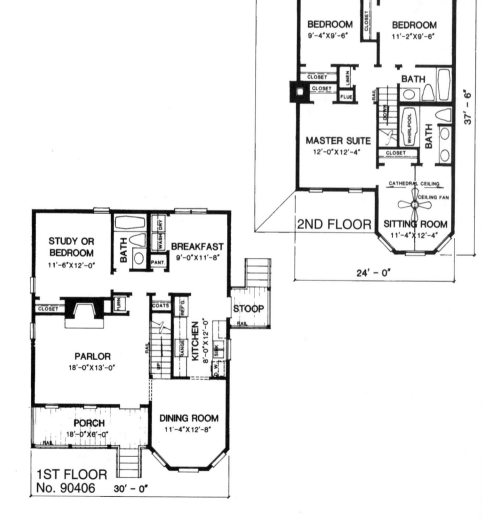

To order your Blueprints, call 1-800-235-5700

■ When ordering this plan— please specify a basement, slab or crawlspace

FIRST FLOOR — 1,128 SQ. FT.
SECOND FLOOR — 1,180 SQ. FT.
BONUS ROOM — 254 SQ. FT.
BASEMENT — 1,114 SQ. FT.
GARAGE — 484 SQ. FT.
DECK — 192 SQ. FT.

TOTAL LIVING AREA:
2,308 SQ. FT.

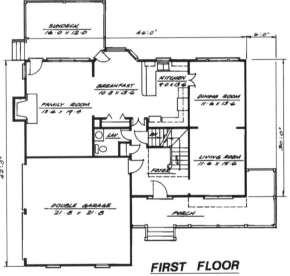

FIRST FLOOR

No. 93237

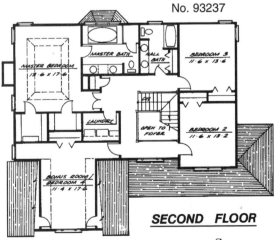

SECOND FLOOR

Country Living with Room to Grow

■ This plan features:

— Three bedrooms

— Two full and one half baths

■ A formal Living Room adjoining the Dining Room, perfectly arranged for entertaining

■ A well-appointed Kitchen with an abundance of work space and a peninsula counter/eating bar, efficiently serving the Dining Room and the Breakfast Bay

■ A spacious Family Room enhanced by a cozy fireplace

■ A second floor Master Suite with a decorative ceiling adding architectural interest, a private bath and two walk-in closets

■ Two additional bedrooms sharing use of the full bath in the hall

■ No materials list is available for this plan

An EXCLUSIVE DESIGN
By Jannis Vann & Associates, Inc.

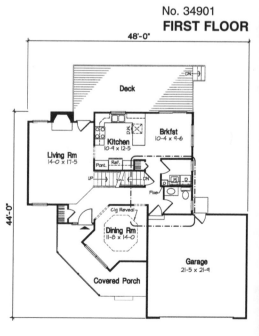

Covered Porch on Farm Style Traditional

- This plan features:
- — Three bedrooms
- — Two and one half baths
- A Dining Room with bay window and elevated ceiling
- A Living Room complete with gas light fireplace
- A two-car Garage
- Ample storage space throughout the home

FIRST FLOOR — 909 SQ. FT
SECOND FLOOR — 854 SQ. FT.
BASEMENT — 899 SQ. FT.
GARAGE — 491 SQ. FT.

TOTAL LIVING AREA:
1,763 SQ. FT.

SECOND FLOOR

Master Br
14-3 x 17-5

Br 3
12-2 x 10-1

Br 2
13-11 x 11-9

Second Floor

Line of Floor Below

Railing

Flue

DN

Opt. Slab/ Crawl Space

Furn.

An
EXCLUSIVE DESIGN
By Karl Kreeger

No. 34901
FIRST FLOOR

48'-0"

44'-0"

Deck

Living Rm
14-0 x 17-5

Kitchen
10-4 x 12-5

Brkfst
10-4 x 9-6

Pant. Ref.

UP

DN

Flue

Clg Reveal

Dining Rm
11-8 x 14-0

Garage
21-5 x 21-4

Covered Porch

One-Story Home Brimming With Amenities

■ This plan features:

— Three bedrooms

— Two full baths

■ Pleasant country look with double dormer windows and wrap around Porch

■ Foyer opens to Dining Room and Activity Room enhanced by tray ceiling, corner fireplace and Sun Deck access

■ Kitchen/Breakfast Room topped by a sloped ceiling, offers an angular serving counter and lots of storage space

■ Secluded Master Bedroom graced with twin walk-in closets and a garden tub bath

■ Two additional bedrooms with easy access to full bath

MAIN FLOOR — 2,079 SQ. FT.
BASEMENT — 2,079 SQ. FT.
GARAGE — 438 SQ. FT.

WHEELCHAIR ACCESSIBLE DETAILS FURNISHED

WHEELCHAIR BATH (OPT.)

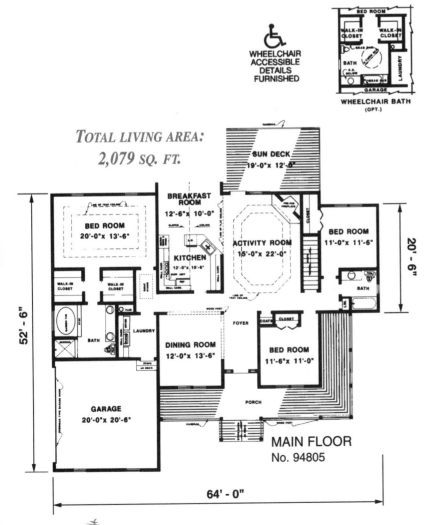

TOTAL LIVING AREA:
2,079 SQ. FT.

MAIN FLOOR
No. 94805

Refer to **Pricing Schedule F** on the order form for pricing information

Veranda Mirrors Two-Story Bay

■ This plan features:

— Four bedrooms

— Two full and one half baths

■ A huge foyer flanked by the formal Parlor and Dining Room

■ An island Kitchen with an adjoining pantry

■ A Breakfast bay and sunken Gathering Room located at the rear of the home

■ Double doors opening to the Master Suite and the book-lined Master Retreat

■ An elegant Master Bath including a raised tub and adjoining cedar closet

FIRST FLOOR — 2,108 SQ. FT.
SECOND FLOOR — 2,109 SQ. FT.
BASEMENT — 1,946 SQ. FT.
GARAGE — 764 SQ. FT.

TOTAL LIVING AREA:
4,217 SQ. FT.

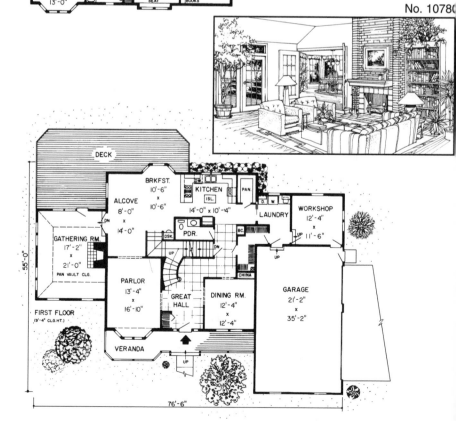

No. 10780

To order your Blueprints, call 1-800-235-5700

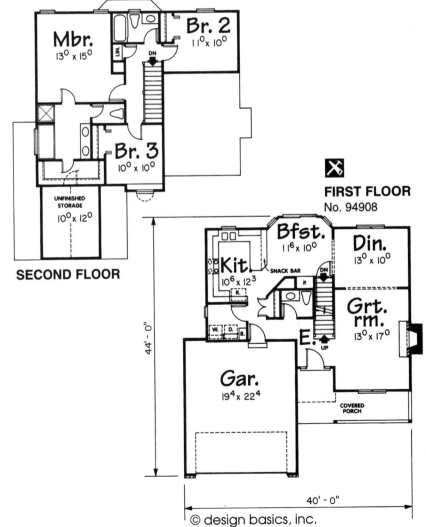

SECOND FLOOR

Mbr.
13⁰ x 15⁰

Br. 2
11⁰ x 10⁰

Br. 3
10⁰ x 10⁰

UNFINISHED STORAGE
10⁰ x 12⁰

FIRST FLOOR
No. 94908

Bfst.
11⁶ x 10⁰

Kit.
10⁶ x 12³

SNACK BAR

Din.
13⁰ x 10⁰

Grt. rm.
13⁰ x 17⁰

Gar.
19⁴ x 22⁴

COVERED PORCH

44' - 0"

40' - 0"

© design basics, inc.

Charming Gabled Porch

■ This plan features:

— Three bedrooms

— Two full and one half baths

■ Formal Dining Room expands into the Great Room for easy entertaining

■ Kitchen snack bar and Breakfast alcove provide two informal eating options

■ A corner Master Bedroom suite has a double vanity bath, a large walk-in closet and an Unfinished Storage area beyond

■ Two additional bedrooms share a full hall bath and linen closet

FIRST FLOOR — 862 SQ. FT.
SECOND FLOOR — 780 SQ. FT.
BASEMENT — 862 SQ. FT.
GARAGE — 454 SQ. FT.

TOTAL LIVING AREA:
1,642 SQ. FT.

Refer to **Pricing Schedule B** on
the order form for pricing information

© 1997 Donald A. Gardner Architects, Inc.

Illusion of Spaciousness

■ This plan features:

— Three bedrooms

— Two full baths

■ Open living spaces and vaulted ceilings create an illusion of spaciousness

■ Cathedral ceilings maximize space in Great Room and Dining Room

■ Kitchen features skylight and breakfast bar

■ Well equipped Master Suite in rear for privacy

■ Two additional bedrooms in front share a full bath

MAIN FLOOR — 1,246 SQ. FT.
GARAGE — 420 SQ. FT.

TOTAL LIVING AREA:
1,246 SQ. FT.

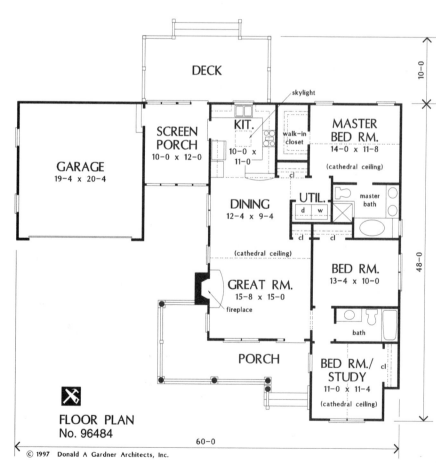

DECK

GARAGE
19-4 x 20-4

SCREEN PORCH
10-0 x 12-0

KIT.
10-0 x 11-0

skylight

walk-in closet

MASTER BED RM.
14-0 x 11-8
(cathedral ceiling)

master bath

DINING
12-4 x 9-4

UTIL.
d w

cl

cl

cl

(cathedral ceiling)

GREAT RM.
15-8 x 15-0
fireplace

BED RM.
13-4 x 10-0

bath

PORCH

BED RM./
STUDY
11-0 x 11-4
(cathedral ceiling)

10-0

48-0

60-0

FLOOR PLAN
No. 96484

© 1997 Donald A Gardner Architects, Inc.

Refer to **Pricing Schedule C** on the order form for pricing information

An EXCLUSIVE DESIGN
By Westhome Planners, Ltd.

SECOND FLOOR

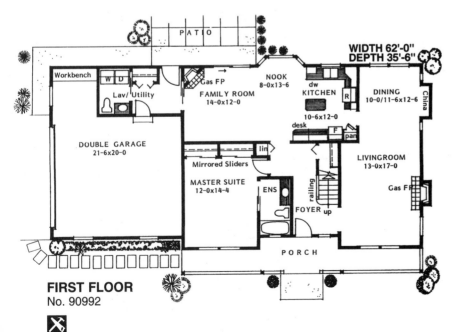

FIRST FLOOR — 1,306 SQ. FT.
SECOND FLOOR — 647 SQ. FT.
GARAGE — 504 SQ. FT.

TOTAL LIVING AREA:
1,953 SQ. FT.

FIRST FLOOR
No. 90992

A Main Floor Master Retreat

■ This plan features:

— Three bedrooms

— Two full and one half baths

■ An island Kitchen with a built-in pantry, desk, double sinks, as well as ample cabinet and counter space

■ A bayed Nook area for informal eating

■ A corner gas fireplace in the Family Room for that cozy touch

■ A built-in china cabinet in the Dining Room

■ A second gas fireplace in the expansive Living Room

■ First floor Master Bedroom with a private Master Bath

■ Two additional bedrooms, accented by dormers and window seats share a double vanity full bath

■ Double garage has a workbench and an entry to both the Utility Room and the Lavatory

Refer to **Pricing Schedule D** on the order form for pricing information

Comfortable Country Porch

■ This plan features:

— Four bedrooms

— Two full and one half baths

■ Wrap-around Porch leads into two-story Foyer with a lovely banister staircase

■ Formal Living Room with arched opening into Family Room with hearth fireplace

■ Efficient, U-shaped Kitchen with a pantry, peninsula serving counter, bright Breakfast area, Garage entry and laundry facilities

■ Corner Master Bedroom offers two closets, a double vanity and jacuzzi

■ Three additional bedrooms convenient to a full bath

■ No materials list available

FIRST FLOOR — 1,348 SQ. FT.
SECOND FLOOR — 1,137 SQ. FT.
BASEMENT — 1,348 SQ. FT.

WIDTH 69'- 0"
DEPTH 37'- 0"

SECOND FLOOR

TOTAL LIVING AREA:
2,485 SQ. FT.

No. 99056

FIRST FLOOR

Refer to **Pricing Schedule G** on the order form for pricing information

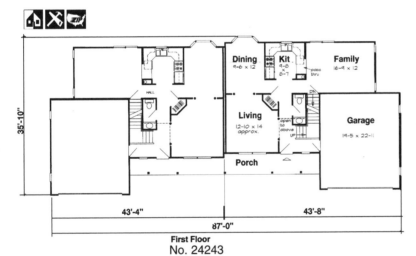

First Floor
No. 24243

43'-4" 43'-8"
87'-0"
35'-10"

Dining 9-6 × 12
Kit 9-8 × 8-7
Family 16-4 × 12
HALL
Living 12-10 × 14 approx.
open to above
UP / DN
pass thru
Garage 19-5 × 22-11
Porch

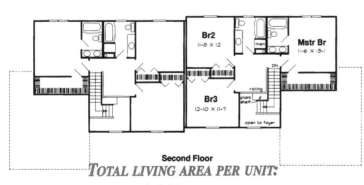

Second Floor

Br2 11-8 × 12
linen
Mstr Br 11-6 × 13-1
Br3 12-10 × 11-7
railing
plant shelf
open to foyer
DN

TOTAL LIVING AREA PER UNIT:
1,644 SQ. FT.
TOTAL LIVING AREA:
3,288 SQ. FT.

Attractive, Affordable Duplex

■ This plan features per unit:

— Three bedrooms

— Two full and a half baths

■ A welcoming front porch sheltering the front entrance

■ A formal Living and Dining Room for ease in entertaining

■ An expansive Family Room that views the efficient Kitchen insuring

■ Uninterupted conversation during meal preparation

■ A large Master Suite with his and her closets and a private Master Bath

■ Two additional bedrooms that share a full hall bath

FIRST FLOOR(PER UNIT) — 819 SQ. FT.
SECOND FLOOR(PER UNIT) — 825 SQ. FT.
GARAGE(PER UNIT) — 470 SQ. FT.
BASEMENT(PER UNIT) — 819 SQ. FT.

Refer to **Pricing Schedule B** on the order form for pricing information

Abundance of Closet Space

An
EXCLUSIVE DESIGN
By Karl Kreeger

■ This plan features:

— Three bedrooms

— Two full baths

■ Roomy walk-in closets in all the bedrooms

■ A Master Bedroom with a decorative ceiling and a private full bath

■ A fireplaced Living Room with sloped ceilings and sliders to the deck

■ An efficient Kitchen, with plenty of cupboard space and a pantry

MAIN AREA —1,532 SQ. FT.
GARAGE — 484 SQ. FT.

TOTAL LIVING AREA:
1,532 SQ. FT.

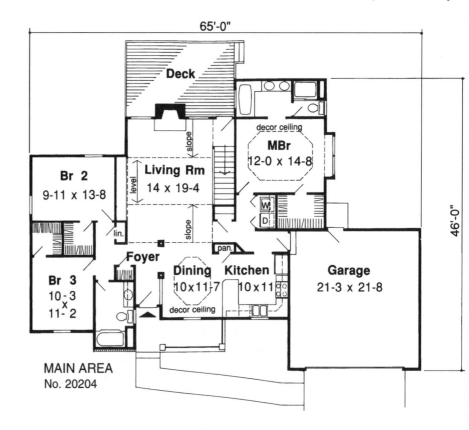

65'-0"

Deck

Br 2
9-11 x 13-8

Living Rm
14 x 19-4

decor ceiling

MBr
12-0 x 14-8

slope

level

slope

W
D

lin.

pan.

Foyer

Br 3
10-3
x
11-2

Dining
10x11-7

decor ceiling

Kitchen
10x11

Garage
21-3 x 21-8

46'-0"

MAIN AREA
No. 20204

To order your Blueprints, call 1-800-235-5700

WIDTH 58'-0"
DEPTH 44'-0"

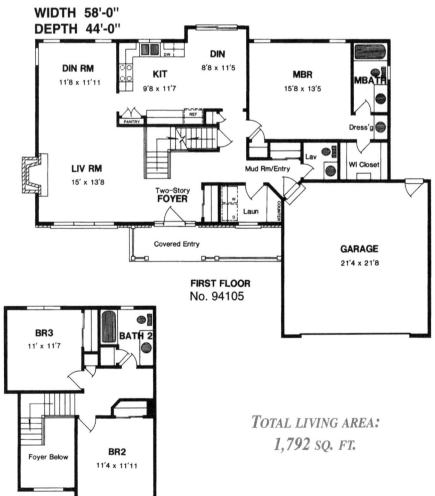

DIN RM
11'8 x 11'11

KIT
9'8 x 11'7

DIN
8'8 x 11'5

MBR
15'8 x 13'5

MBATH

PANTRY

REF

DW

Dress'g

LIV RM
15' x 13'8

Lav

WI Closet

Mud Rm/Entry

Two-Story
FOYER

W
D

Laun

COUNTER

GARAGE
21'4 x 21'8

Covered Entry

FIRST FLOOR
No. 94105

BR3
11' x 11'7

BATH 2

Foyer Below

BR2
11'4 x 11'11

SECOND FLOOR

TOTAL LIVING AREA:
1,792 SQ. FT.

Classic Style and Comfort

◼ This plan features:

— Three bedrooms

— Two full and one half bath

◼ Covered Entry leads into two-story Foyer with a dramatic landing staircase brightened by decorative window

◼ Spacious Living/Dining Room combination with hearth fireplace and decorative windows

◼ Hub Kitchen with built-in pantry and informal Dining area with sliding glass door to rear yard

◼ First floor Master Bedroom offers a walk-in closet, dressing area and full bath

◼ Two additional bedrooms on second floor share a full bath

◼ No materials list available

FIRST FLOOR — 1,281 SQ. FT.
SECOND FLOOR —511 SQ. FT.
GARAGE — 467 SQ. FT.

Refer to **Pricing Schedule C** on the order form for pricing information

Moderate Ranch Has Features of Much Larger Plan

■ This plan features:

— Three bedrooms

— Two full baths

■ A large Great Room with a vaulted ceiling and a stone fireplace with bookshelves on either side

■ A spacious Kitchen with ample cabinet space conveniently located next to the large Dining Room

■ A Master Suite having a large bath with a garden tub, double vanity and a walk-in closet

■ Two other large bedrooms, each with a walk-in closet and access to the full bath

■ This plan is available with a basement, slab or crawl space foundation — please specify

MAIN FLOOR — 1,811 SQ. FT.

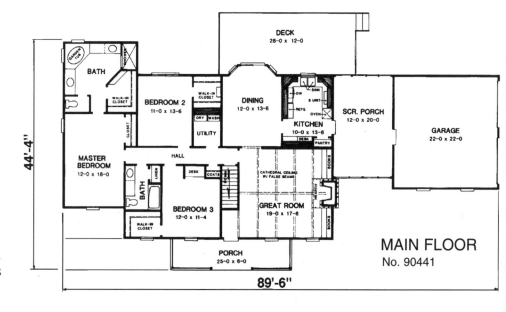

MAIN FLOOR
No. 90441

TOTAL LIVING AREA:
1,811 SQ. FT.

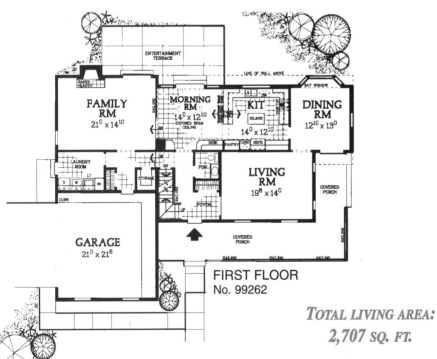

FIRST FLOOR
No. 99262

TOTAL LIVING AREA:
2,707 SQ. FT.

WIDTH 63'-0"
DEPTH 48'-0"

SECOND FLOOR

Farmhouse Favorite

■ This plan features:

— Four bedrooms

— Two full and one half baths

■ Covered Porch adds living space outdoors and shelters entrance into Foyer

■ Formal Living Room highlighted by full length windows and open to formal Dining Room with bay window and Porch access

■ Spacious Kitchen with work island, pass-thru to Morning Room and Terrace, and nearby Laundry/Garage entry

■ Comfortable Family Room with a raised hearth fireplace and access to Entertainment Terrace

■ Corner Master Bedroom offers two closets and a pampering bath

■ Two or three additional bedrooms with ample closets, share a double vanity bath and attic storage

FIRST FLOOR — 1,595 SQ. FT.
SECOND FLOOR — 1,112 SQ. FT.

Refer to **Pricing Schedule D** on the order form for pricing information

Eye-Appealing Balance

■ This plan features:

— Three bedrooms

— Two full and one half baths

■ Arched Portico enhances entry into Gallery and spacious Living Room, with focal point fireplace surrounded by glass

■ Cathedral ceilings top Family Room and formal Dining Room

■ An efficient Kitchen with breakfast area opens to Family Room, Utility Room with convenient Garage entry

■ Corner Master Bedroom suite with access to covered Patio and private bath with a double vanity and garden window tub

■ Two additional bedrooms with walk-in closets share a full bath

■ No materials list available

MAIN FLOOR — 2,470 SQ. FT.
GARAGE — 483 SQ. FT.

TOTAL LIVING AREA:
2,470 SQ. FT.

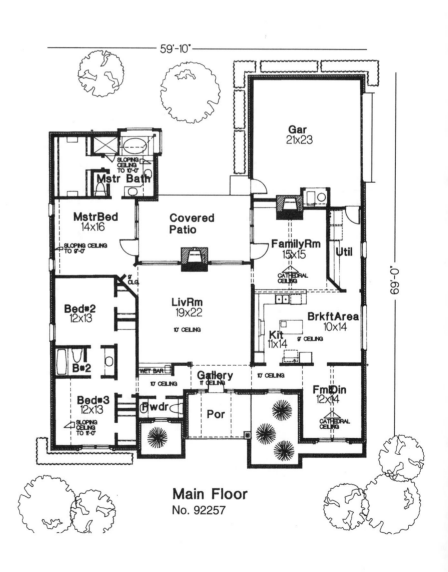

Main Floor
No. 92257

136

An
EXCLUSIVE DESIGN
By *Jannis Vann & Associates, Inc.*

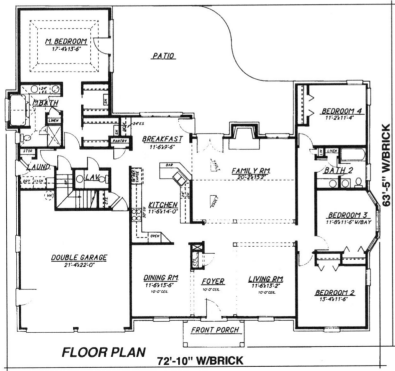

FLOOR PLAN

72'-10" W/BRICK

63'-5" W/BRICK

No. 93253

Today's Lifestyle

■ This plan features:

— Four bedrooms

— Two full and one half baths

■ A large Family Room with a fireplace and access to the patio

■ A Breakfast Area that flows directly into the Family Room

■ A well-appointed Kitchen equipped with an eating bar, double sinks, built-in pantry and an abundance of counter and cabinet space

■ A Master Suite with a decorative ceiling and a private Bath

■ Three additional bedrooms that share a full bath

MAIN AREA — 2,542 SQ. FT.
GARAGE — 510 SQ. FT.

TOTAL LIVING AREA:
2,542 SQ. FT.

Refer to **Pricing Schedule C** on the order form for pricing information

Country Living in Any Neighborhood

- This plan features:
- — Three bedrooms
- — Two full and two half baths
- An expansive Family Room with fireplace
- A Dining Room and Breakfast Nook lit by flowing natural light from bay windows
- A first floor Master Suite with a double vanitied bath that wraps around his-n-her closets
- Optional basement, slab or crawl space foundation — please specify when ordering

FIRST FLOOR — 1,477 SQ. FT.
SECOND FLOOR — 704 SQ. FT.
BASEMENT — 1,374 SQ. FT.

**TOTAL LIVING AREA:
2,181 SQ. FT.**

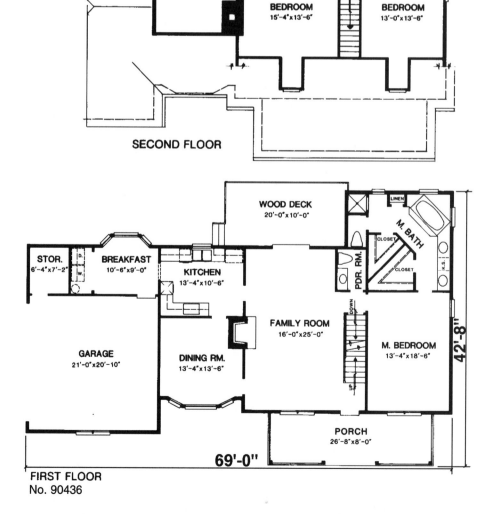

SECOND FLOOR

CLOSET
DRESS. BATH DRESS.
CLOSET

STORAGE
13'-0"x9'-6"

BEDROOM
15'-4"x13'-6"

DOWN

BEDROOM
13'-0"x13'-6"

WOOD DECK
20'-0"x10'-0"

LINEN

M. BATH

CLOSET

PDR. RM.

CLOSET

STOR.
6'-4"x7'-2"

BREAKFAST
10'-6"x9'-0"

KITCHEN
13'-4"x10'-6"

FAMILY ROOM
16'-0"x25'-0"

DOWN

M. BEDROOM
13'-4"x18'-6"

42'-8"

GARAGE
21'-0"x20'-10"

DINING RM.
13'-4"x13'-6"

UP

PORCH
26'-8"x8'-0"

69'-0"

FIRST FLOOR
No. 90436

To order your Blueprints, call 1-800-235-5700

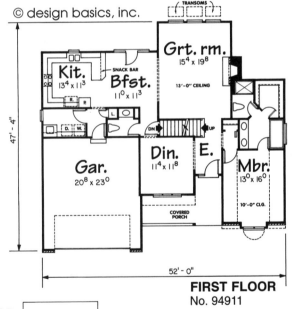

© design basics, inc.

FIRST FLOOR — 1,405 SQ. FT.
SECOND FLOOR — 453 SQ. FT.
BONUS ROOM — 300 SQ. FT.
BASEMENT — 1,405 SQ. FT.

TOTAL LIVING AREA:
1,858 SQ. FT.

Kit.
13⁴ x 11³

Grt. rm.
15⁴ x 19⁸
13'-0" CEILING

SNACK BAR

Bfst.
11⁰ x 11³

Gar.
20⁸ x 23⁰

Din.
11⁴ x 11⁸

E.

Mbr.
13⁰ x 16⁰
10'-0" CLG.

COVERED PORCH

47' - 4"

52' - 0"

FIRST FLOOR
No. 94911

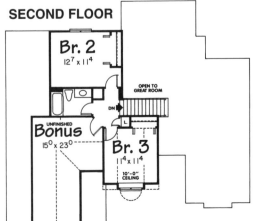

SECOND FLOOR

Br. 2
12⁷ x 11⁴

OPEN TO GREAT ROOM

Bonus
15⁰ x 23⁰
UNFINISHED

Br. 3
11⁴ x 11⁴
10'-0" CEILING

Fieldstone Facade and Arched Windows

■ This plan features:

— Three bedrooms

— Two full baths

■ Inviting Covered Porch shelters entrance

■ Expansive Great Room enhanced by warm fireplace and three transom windows

■ Breakfast area adjoins Great Room giving a feeling of more space

■ An efficient Kitchen with counter snack bar and nearby laundry and Garage entry

■ A first floor Master Bedroom suite with an arched window below a sloped ceiling and a double vanity bath

■ Two additional bedrooms share a Bonus area and a full bath on the second floor

An EXCLUSIVE DESIGN
By Plan One Homes, Inc.

Luxurious Master Suite

■ This plan features:

—Four bedrooms

—Three full and one half baths

■ Fireplace in the formal Living Room and one in the Family Room

■ Optional Sun Room expanding living space

■ Kitchen located between the Breakfast Area and the formal Dining Room

■ A plush bath with a whirlpool tub, vaulted ceiling over the bath and a tray ceiling over the bedroom highlighting the Master Suite

■ Two additional bedrooms, each with private access to full bath

■ Bonus room available for future expansion

FIRST FLOOR — 1,523 SQ. FT.
SECOND FLOOR — 1,370 SQ. FT.
BASEMENT — 1,722 SQ. FT.
GARAGE — 484 SQ. FT.

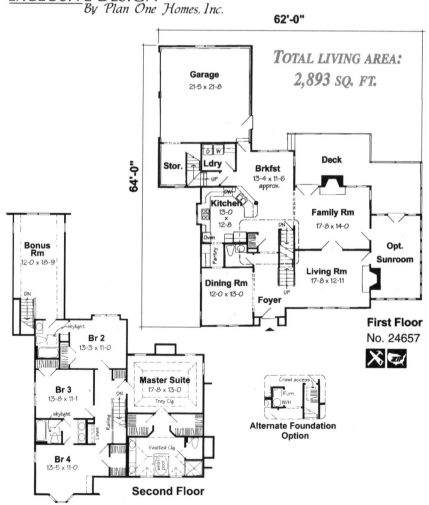

TOTAL LIVING AREA: 2,893 SQ. FT.

Garage
21-5 x 21-8

Stor. Ldry

Brkfst
13-4 x 11-6
approx.

Deck

Kitchen
13-0 x 12-8

Family Rm
17-8 x 14-0

Opt. Sunroom

Living Rm
17-8 x 12-11

Dining Rm
12-0 x 13-0

Foyer

First Floor
No. 24657

Bonus Rm
12-0 x 18-9

Br 2
13-3 x 11-0

Br 3
13-8 x 11-1

Master Suite
17-8 x 13-0
Trey Clg

Br 4
13-5 x 11-0

Vaulted Clg

Second Floor

Crawl access
Furn.
W/H

Alternate Foundation Option

62'-0"

64'-0"

To order your Blueprints, call 1-800-235-5700

© 1992 Donald A. Gardner Architects, Inc.

SECOND FLOOR PLAN

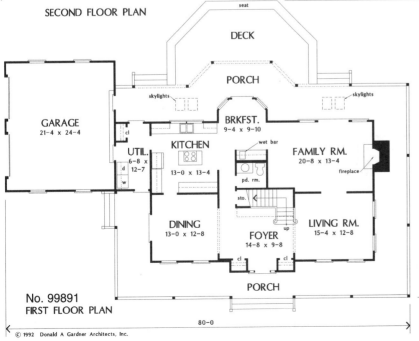

No. 99891
FIRST FLOOR PLAN

© 1992 Donald A Gardner Architects, Inc.

Grand Four Bedroom Farmhouse

■ This plan features:

— Four bedrooms

— Two full and one half baths

■ Double gables, wrap-around Porch and custom window details add farmhouse appeal

■ Formal Living and Dining rooms connected by Foyer in front, while casual living areas are in the rear

■ Efficient Kitchen with island cooktop and easy access to all eating areas

■ Spacious Master bedroom suite features walk-in closet and pampering bath

FIRST FLOOR — 1,357 SQ. FT.
SECOND FLOOR — 1,204 SQ. FT.
GARAGE & STORAGE — 546 SQ. FT.

TOTAL LIVING AREA:
2,561 SQ. FT.

Refer to **Pricing Schedule D** on the order form for pricing information

Country Style for Today

■ This plan features:

— Four bedrooms

— Two full and one half baths

■ Two bay windows in the formal Living Room with a heat-circulating fireplace to enhance the mood and warmth

■ A spacious formal Dining Room with a bay window and easy access to the Kitchen

■ An octagon-shaped Dinette defined by columns, dropped beams and a bay window

■ An efficient island Kitchen with ample storage and counter space

■ A Master Suite equipped with a large whirlpool tub and double vanity

■ Three additional bedrooms that share a full hall bath

FIRST FLOOR — 1,132 SQ. FT.
SECOND FLOOR — 1,020 SQ. FT.
BASEMENT — 1,026 SQ. FT.
GARAGE & STORAGE — 469 SQ. FT.
LAUNDRY/MUDROOM — 60 SQ. FT.

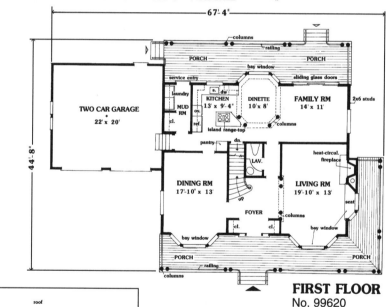

FIRST FLOOR
No. 99620

TOTAL LIVING AREA:
2,212 SQ. FT.

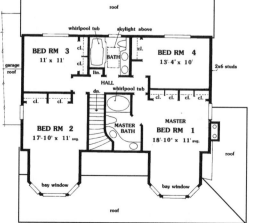

SECOND FLOOR

142

To order your Blueprints, call 1-800-235-5700

FIRST FLOOR PLAN
No. 93306

- DINETTE 10 X 8'4
- D. W.
- LAUND.
- STEP
- B.C.
- P.R.
- KITCHEN 10 X 18'6
- DW
- FAMILY RM. 16 X 13
- DN
- UP
- FLR. ABV.
- GARAGE 20'6 X 22
- DINING RM. 10'6 X 11'6
- FOYER HIGH CLG.
- LIVING RM. 10'6 X 11'6
- 16ft. DOOR
- PORCH
- STEP
- 50'0"
- 31'0"

TOTAL LIVING AREA:
1,672 SQ. FT.

SECOND FLOOR PLAN

- 60x36 whirlpool
- 36x36 shwr
- M.BATH
- M.B.R. 13 X 16'8
- twl
- 6sh
- twl
- BATH 2
- DN
- 6SH
- ROOF
- BALCONY
- RAILING
- B.R.#3 10'6 X 11'6
- FOYER, BELOW
- B.R.#2 10'6 X 11'6

Cozy and Comfortable

■ This plan features:

— Three bedrooms

— Two full and one half baths

■ Center Foyer leads into formal Living and Dining rooms

■ Open Family Room accented by hearth fireplace

■ Efficient Kitchen with peninsula counter, nearby Laundry and Garage entry, and Dinette with access to rear yard

■ Corner Master Bedroom offers a plush bath with a double vanity and whirlpool tub

■ Two additional bedrooms with ample closets share a full bath

■ No materials list is available for this plan

FIRST FLOOR — 884 SQ. FT.
SECOND FLOOR — 788 SQ. FT.
BASEMENT — 884 SQ. FT.
GARAGE — 450 SQ. FT.

An
EXCLUSIVE DESIGN
By Patrick Morabito, A.I.A. Architect

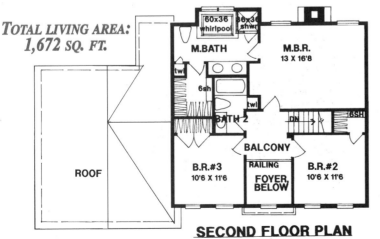

Refer to **Pricing Schedule C** on the order form for pricing information

One-of-a-Kind

■ This plan features:

— Three bedrooms

— Two full and a half baths

■ A porch sheltering the entry

■ A fireplaced Dining Room with warmth and atmosphere

■ A corner fireplace adding a focal point to the Parlor

■ An island Kitchen with a Breakfast area and walk-in pantry

FIRST FLOOR — 955 SQ. FT.

SECOND FLOOR — 864 SQ. FT.

BASEMENT — 942 SQ. FT.

TOTAL LIVING AREA: 1,819 SQ. FT.

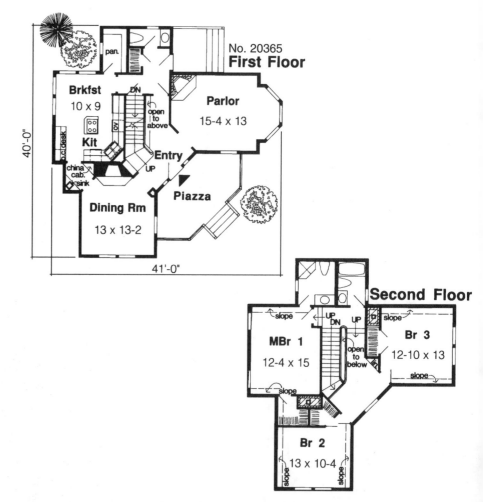

No. 20365
First Floor

Brkfst 10 x 9

Kit

Parlor 15-4 x 13

Entry

Piazza

Dining Rm 13 x 13-2

40'-0"

41'-0"

Second Floor

MBr 1 12-4 x 15

Br 3 12-10 x 13

Br 2 13 x 10-4

To order your Blueprints, call 1-800-235-5700

Refer to **Pricing Schedule C** on the order form for pricing information

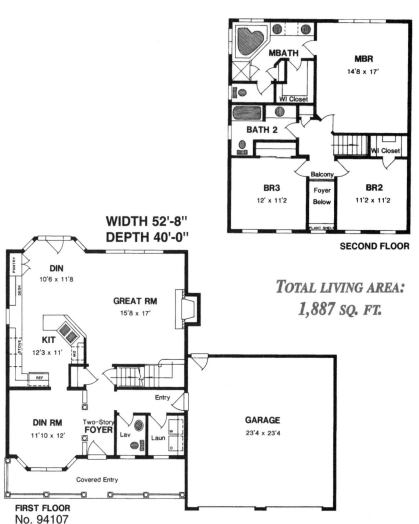

MBATH

MBR
14'8 x 17'

WI Closet

BATH 2

WI Closet

BR3
12' x 11'2

Balcony
Foyer
Below

BR2
11'2 x 11'2

PLANT SHELF

SECOND FLOOR

WIDTH 52'-8"
DEPTH 40'-0"

PANTRY

DESK

DIN
10'6 x 11'8

GREAT RM
15'8 x 17'

STOVE

KIT
12'3 x 11'

REF

DIN RM
11'10 x 12'

Two-Story
FOYER

Lav

Laun
W D

Entry

GARAGE
23'4 x 23'4

Covered Entry

FIRST FLOOR
No. 94107

TOTAL LIVING AREA:
1,887 SQ. FT.

Spacious Country Charm

■ This plan features

— Three Bedrooms

— Two full and one half baths

■ Comfortable front Porch leads into bright, two-story Foyer

■ Pillars frame entrance to formal Dining Room highlighted by bay window

■ Expansive Great Room accented by hearth fireplace and triple window opens to Kitchen/Dining area

■ Efficient Kitchen with loads of counter space and Dining area with access to rear yard

■ Corner Master Bedroom offers triple windows and a luxurious Master Bath

■ Two additional bedrooms, with ample closets, share a full bath

■ No materials list available

FIRST FLOOR — 961 SQ. FT.
SECOND FLOOR — 926 SQ. FT.
GARAGE — 548 SQ. FT.

Refer to **Pricing Schedule C** on the order form for pricing information

Warm Country Home

■ This plan features:

— Three bedrooms

— Two full and one half baths

■ A wrap-around Porch extends living outdoors

■ Expansive Activity Room with a huge fireplace separated from Dining Room by a lovely banister staircase

■ Efficient Kitchen with work island/serving counter for Breakfast Room, Laundry and Garage entry

■ First floor Master Bedroom with a tray ceiling and a pampering bath

■ Two second floor bedrooms with dormer windows share a full bath

FIRST FLOOR — 1,531 SQ. FT.
SECOND FLOOR — 640 SQ. FT.

TOTAL LIVING AREA:
2,171 SQ. FT.

SECOND FLOOR PLAN

No. 94806

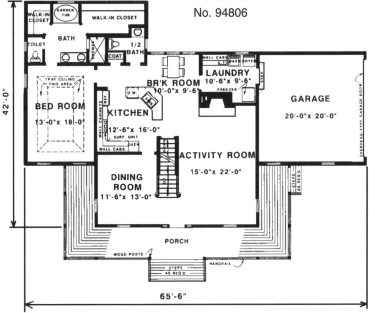

FIRST FLOOR PLAN

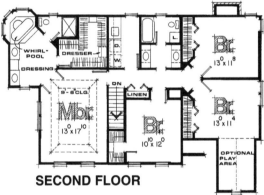

SECOND FLOOR

FIRST FLOOR — 1,322 SQ. FT.
SECOND FLOOR — 1,272 SQ. FT.
BASEMENT — 1,322 SQ. FT.
GARAGE — 468 SQ. FT.

TOTAL LIVING AREA:
2,594 SQ. FT.

FIRST FLOOR
No. 94932

© design basics, inc.

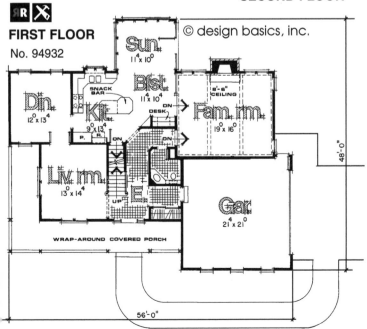

WRAP-AROUND COVERED PORCH

56'-0"

48'-0"

Welcoming Wrap-Around Porch

■ This plan features:

— Four bedrooms

— Two full and one half baths

■ Front porch expands living space and accesses easy-care tile Entry

■ Formal Living and Dining rooms connected by French doors

■ An efficient Kitchen with serving counter/snackbar, pantry, Breakfast area and Sunroom beyond

■ Sunken Family Room with beamed ceiling and inviting fireplace

■ Corner Master Bedroom with decorative ceiling and French doors into spacious dressing area with a whirlpool window tub and large walk-in closet

■ Three additional bedrooms, one with an optional play area, share a double vanity bath

Refer to **Pricing Schedule D** on the order form for pricing information

©1996 Donald A. Gardner Architects, Inc.   B. NATHAN

Not a Typical Farmhouse

■ This plan features:

— Four bedrooms

— Two full and one half baths

■ A large center gable with a palladian window and a gently vaulted portico

■ Formal Dining Room and a Living Room/study highlighted by tray ceilings

■ Spacious Family Room, adjoining the Living Room and the Breakfast area

■ Nine foot ceilings adding volume to the first floor

■ Efficient Kitchen with a center work island and a roomy pantry

■ Gracious Master Suite, topped by a tray ceiling, contains a generous walk-in closet and a skylit bath with a double vanity, linen closet and a garden tub

■ Three additional bedrooms share a full bath

SECOND FLOOR PLAN

TOTAL LIVING AREA:
2,475 SQ. FT.

FIRST FLOOR — 1,299 SQ. FT.
SECOND FLOOR — 1,176 SQ. FT.
GARAGE & STORAGE — 641 SQ. FT.
BONUS ROOM — 464 SQ. FT.

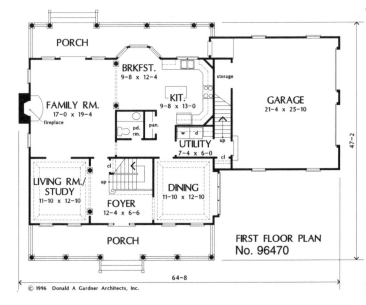

FIRST FLOOR PLAN
No. 96470

© 1996 Donald A Gardner Architects, Inc.

To order your Blueprints, call 1-800-235-5700

■ No materials list is available for this plan

■ When ordering this plan—please specify basement   or crawl space foundation

M·B

B·2

BEDRM • 3
13/10 x 10/8

W·I·C

BONUS RM
19/6 x 13/0

MASTER BEDROOM
13/2 x 14/8

DN

OPEN TO BELOW

BEDRM • 2
13/2 x 12/4

UPPER FLOOR
No. 91073

TOTAL LIVING AREA:
2,209 SQ. FT.

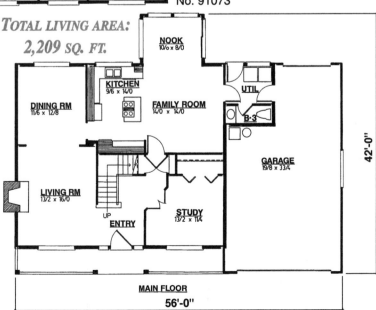

NOOK
10/o x 8/0

KITCHEN
9/6 x 14/0

UTIL

DINING RM
11/6 x 12/8

FAMILY ROOM
14/0 x 14/0

B·3

GARAGE
19/8 x 33/4

42'-0"

LIVING RM
13/2 x 16/0

UP

ENTRY

STUDY
13/2 x 11/4

MAIN FLOOR
56'-0"

Enjoy a Summer Breeze on the Covered Porch

■ This plan features:

— Three bedrooms

— Two and one half baths

■ A large Living Room enhanced by a fireplace and an open entry to the formal Dining Room

■ An efficient L-shaped Kitchen with cooktop island and open lay-out to the Family Room and Nook area creating a feeling of spaciousness

■ Two generously sized bedrooms that share a full bath

■ A Master Suite that includes a walk-in closet and spa tub with garden windows

■ A large Bonus Room for you to decide on

MAIN FLOOR — 1,240 SQ. FT.
UPPER FLOOR — 969 SQ. FT.
BONUS ROOM — 254 SQ. FT.
GARAGE — 550 SQ. FT.

Brick and Wood Highlighted by Sunbursts

■ This plan features:

— Three bedrooms

— Two full and one half baths

■ Sheltered Porch entrance leads into two-story Foyer

■ Beautiful bay window brightens Living Room which opens into the formal Dining Room

■ Efficient, L-shaped Kitchen has a built-in pantry and Dining area

■ Comfortable Family Room with focal point fireplace topped by cathedral ceiling

■ Master Bedroom offers two closets and a private bath

■ Two additional bedrooms with large closets share a full bath

■ No materials list available

FIRST FLOOR — 1,094 SQ. FT.
SECOND FLOOR — 719 SQ. FT.
GARAGE — 432 SQ. FT.

TOTAL LIVING AREA:
1,813 SQ. FT.

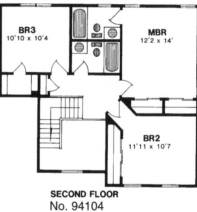

BR3
10'10 x 10'4

MBR
12'2 x 14'

BR2
11'11 x 10'7

SECOND FLOOR
No. 94104

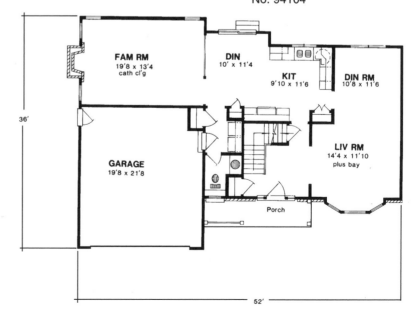

FAM RM
19'8 x 13'4
cath cl'g

DIN
10' x 11'4

KIT
9'10 x 11'6

DIN RM
10'8 x 11'6

LIV RM
14'4 x 11'10
plus bay

GARAGE
19'8 x 21'8

Porch

36'

52'

FIRST FLOOR

To order your Blueprints, call 1-800-235-5700

Refer to **Pricing Schedule B** on the order form for pricing information

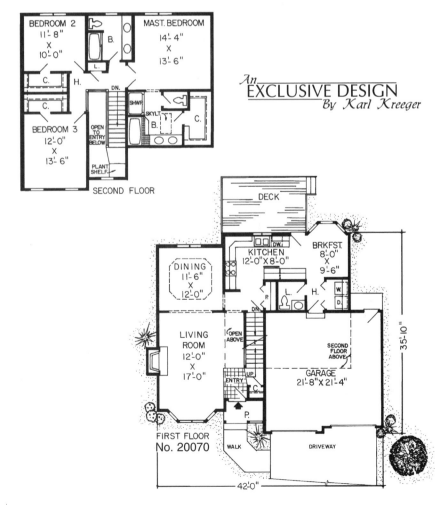

BEDROOM 2
11'-8"
X
10'-0"

MAST. BEDROOM
14'-4"
X
13'-6"

B.

L.

C.

C.

H.

DN.

SHWR

SKYLT

B.

C.

BEDROOM 3
12'-0"
X
13'-6"

OPEN TO ENTRY BELOW

PLANT SHELF

SECOND FLOOR

An
EXCLUSIVE DESIGN
By Karl Kreeger

DECK

KITCHEN
12'-0" X 8'-0"

DW.

BRKFST.
8'-0"
X
9'-6"

DINING
11'-6"
X
12'-0"

P.

L.

H.

W.

D.

DN.

LIVING ROOM
12'-0"
X
17'-0"

OPEN ABOVE

SECOND FLOOR ABOVE

35'-10"

UP

ENTRY

C.

GARAGE
21'-8" X 21'-4"

P.

FIRST FLOOR
No. 20070

WALK

DRIVEWAY

42'-0"

Sheltered Porch is an Inviting Entrance

■ This plan features:

— Three bedrooms

— Two full and one half baths

■ A dramatic two-story entry

■ A fireplaced Living Room

■ A modern Kitchen flowing easily into a sunny Breakfast Nook

■ A formal Dining Room with elegant decorative ceiling

■ A Master Bedroom highlighted by a skylit bath

FIRST FLOOR — 877 SQ. FT.
SECOND FLOOR — 910 SQ. FT.
BASEMENT — 877 SQ. FT.
GARAGE — 458 SQ. FT.

TOTAL LIVING AREA:
1,787 SQ. FT.

©1996 Donald A. Gardner Architects, Inc.

B. NATHAN

Easy Living Plan

■ This plan features:

— Three bedrooms

— Two full baths

■ Sunlit Foyer flows easily into the generous Great Room

■ Great Room crowned in a cathedral ceiling and accented by a fireplace

■ Accent columns define the open Kitchen and Breakfast Bay

■ Master Bedroom topped by a tray ceiling and highlighted by a well-appointed Master Bath

■ Two additional bedrooms share a skylit bath in the hall

MAIN AREA — 1,864 SQ. FT.
BONUS ROOM — 319 SQ. FT.
GARAGE — 503 SQ. FT.

TOTAL LIVING AREA:
1,864 SQ. FT.

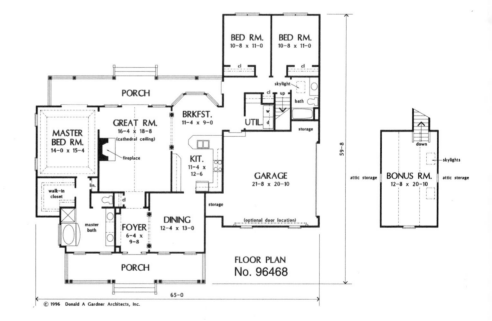

To order your Blueprints, call 1-800-235-5700

Main floor — 3,818 sq. ft.
Garage — 816 sq. ft.

Total living area:
3,818 sq. ft.

Luxurious Masterpiece

■ This plan features:

— Four bedrooms

— Three full and one half baths

■ Expansive formal Living Room with a fourteen foot ceiling and a raised hearth fireplace

■ Informal Family Room offers another fireplace, wet bar, cathedral ceiling and access to the Covered Patio

■ Hub Kitchen with a cooktop island, peninsula counter/snack-bar, and a bright breakfast area

■ French doors lead into a quiet Study offering many uses

■ Private Master Bedroom enhanced by pullman ceiling and lavish his-n-her baths

■ Three additional bedrooms with walk-in closets and private access to a full bath

■ No materials list available

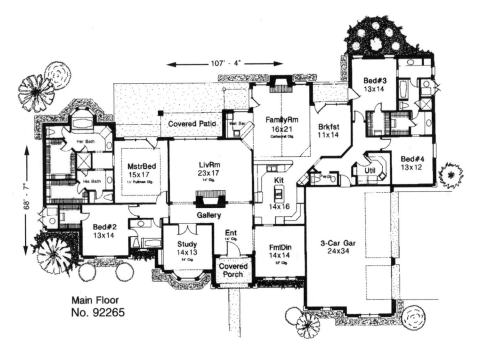

Main Floor
No. 92265

Refer to Pricing Schedule D on the order form for pricing information

Romance Personified

■ This plan features:

— Three bedrooms

— Two full and one half baths

■ A spacious Family Room including a fireplace flanked by bookshelves

■ A sunny Breakfast Bay and adjoining country Kitchen with a peninsula counter

■ An expansive Master Suite spanning the width of the house including built-in shelves, walk-in closet, and a private bath with every amenity

■ A full bath that serves the two other bedrooms tucked into the gables at the front of the house

■ Optional basement or crawl space foundation — please specify when ordering

FIRST FLOOR — 1,366 SQ. FT.
SECOND FLOOR — 1,196 SQ. FT.
BASEMENT — 1,250 SQ. FT.
GARAGE — 484 SQ. FT.

TOTAL LIVING AREA:
2,562 SQ. FT.

SECOND FLOOR

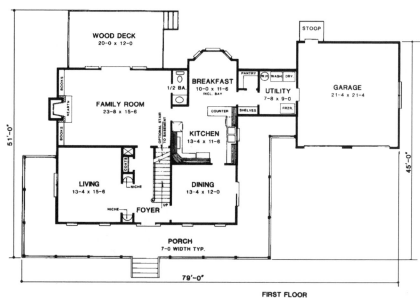

FIRST FLOOR
No. 90439

154 To order your Blueprints, call 1-800-235-5700

Refer to **Pricing Schedule F** on the order form for pricing information

First floor — 2,839 sq. ft.
Second floor — 1,111 sq. ft.
Garage — 885 sq. ft.

Total living area:
3,950 sq. ft.

© design basics, inc.

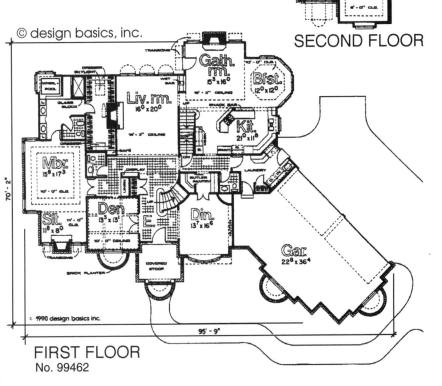

SECOND FLOOR

FIRST FLOOR
No. 99462

Spectacular Two-Story Entry

■ This plan features:

— Four bedrooms

— Two full, two three-quarter and two half baths

■ A spectacular two-story high entry showcasing the floating, curved staircase

■ A butler's pantry for formal serving into the Dining Room

■ French doors into the spider beamed Den

■ Repetitive arched windows at the rear of the Living Room

■ Gathering Room sharing a pass-through wetbar with Living Room

■ Luxurious Master Suite has a tiered ceiling, fireplace and a lavish bath

■ Three secondary bedrooms with boxed ceilings, generous walk-in closets and private baths

■ No materials list available

Refer to **Pricing Schedule A** on the order form for pricing information

Decorative Detailing Adds Charm

■ This plan features:

— Three bedrooms

— One full and one three-quarter baths

■ A Living Room with a cozy fireplace and sloped ceiling

■ An efficient Kitchen equipped with a plant shelf easily accessible to the Dining Room

■ A Master Bedroom with a decorative ceiling and a private bath

■ A second bath equipped with a washer and dryer

MAIN AREA — 1,441 SQ. FT.
GARAGE — 672 SQ. FT.

TOTAL LIVING AREA:
1,441 SQ. FT.

Floor Plan
No. 34005

52'-0"

38'-0"

Patio

slope   slope

plant shelf →
Kitchen
11-8 x 11-4

Living Rm
15-4 x 18

decor. ceiling

MBr 1
13-4 x 13-11

DN

W
D

UP

Dining
11-8 x 13

decor. ceiling

Br 3
10-6 x 11-8

Br 2
11-7 x 11-8

An
EXCLUSIVE DESIGN
By Karl Kreeger

FIRST FLOOR — 2,807 SQ. FT.
SECOND FLOOR — 1,063 SQ. FT.
GARAGE — 633 SQ. FT.

TOTAL LIVING AREA:
3,870 SQ. FT.

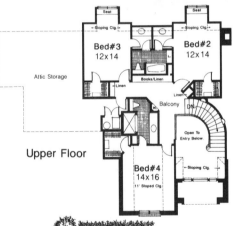

Bed#3
12 x 14

Bed#2
12 x 14

Attic Storage

Books/Linen

Linen

Balcony

Open To
Entry Below

Upper Floor

Bed#4
14 x 16
11' Sloped Clg.

Covered Patio

Brkfst
15 x 14

FamilyRm
19 x 19

3-Car Gar
8'-4" Clg.
28 x 24

Kit
15 x 16

Lania

Gallery

MstrBed
16 x 18

Covered
Porch

Storage

Util

Pantry

Study
16 x 14

FmlDin
12 x 12

Ent

LivRm
14 x 17

Covered
Porch

**Main Floor
No. 92274**

80' - 0"

65' - 4"

Elegant Residence

- This plan features:

— Four bedrooms

— Three full and one half baths

- Two-story glass Entry enhanced by a curved staircase

- Open Living/Dining Room with decorative windows makes entertaining easy

- Large, efficient Kitchen with cooktop/work island, huge walk-in pantry, Breakfast room, butler's pantry and Utility/Garage entry

- Comfortable Family Room with hearth fireplace, built-ins and access to Covered Patio

- Cathedral ceiling tops luxurious Master Bedroom offering a private Lanai, skylit bath, double walk-in closet, and adjoining Study

- Three second floor bedrooms with walk-in closets and private access to full baths

- No materials list available

Refer to **Pricing Schedule C** on
the order form for pricing information

Abundance of Windows for Natural Lighting

■ This plan features:

— Four bedrooms

— Two full and one half baths

■ Ten foot ceiling above transom windows and hearth fireplace accent the Great Room

■ Island counter/snack bar, pantry and desk featured in Kitchen/Breakfast area

■ Kitchen conveniently accesses laundry area and Garage

■ Beautiful arched window under volume ceiling in Bedroom two

■ Master Bedroom suite features decorative ceilings, walk-in closets and double vanity bath with a whirlpool tub

■ Two additional bedrooms with ample closets share a full bath

FIRST FLOOR — 944 SQ. FT.
SECOND FLOOR — 987 SQ. FT.
BASEMENT — 944 SQ. FT.
GARAGE — 557 SQ. FT.

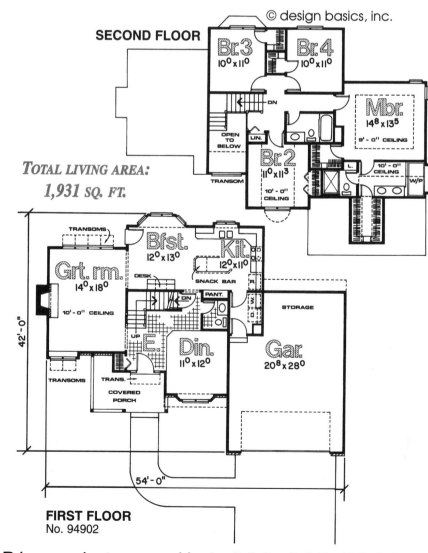

© design basics, inc.

SECOND FLOOR

TOTAL LIVING AREA:
1,931 SQ. FT.

FIRST FLOOR
No. 94902

To order your Blueprints, call 1-800-235-5700

Refer to **Pricing Schedule D** on
the order form for pricing information

Country Style For Today

■ This plan features:

— Three bedrooms

— Two full and one half baths

■ A wide wrap-around porch for a farmhouse style

■ A spacious Living Room with double doors and a large front window

■ A garden window over the double sink in the huge, country Kitchen with two islands, one a butcher block, and the other an eating bar

■ A corner fireplace in the Family Room enjoyed throughout the Nook and Kitchen, thanks to an open layout

■ A Master Suite with a spa tub, and a huge walk-in closet as well as a shower and double vanity

FIRST FLOOR — 1,785 SQ. FT.
SECOND FLOOR — 621 SQ. FT.

TOTAL LIVING AREA:
2,406 SQ. FT.

SECOND FLOOR

DECK

BEDROOM 2
14⁰ x 15⁸

OPEN TO BELOW

BRIDGE
DN

OPEN TO BELOW

WH
LINEN

BEDROOM 3
14⁰ x 11⁸

FIRST FLOOR
No. 91700

55'-0"

DECK

DN

DN

NOOK
12⁶ x 10⁰

FIREPLACE
FAMILY ROOM
21⁰ x 15⁶

MASTER SUITE
14⁰ x 14⁸

REF
ISLAND

KITCHEN
15⁶ x 14⁸

GARDEN WINDOW

BUTCHER BLOCK

WALK-IN CLOSET

SHOWER

FAU

SPA

DW

DESK

UP

R & O

DN

PANTRY

DINING ROOM
11⁶ x 11⁰

LIVING ROOM
14⁰ x 14²

DN

UTILITY

WSH DRY

WH

PORCH

DN

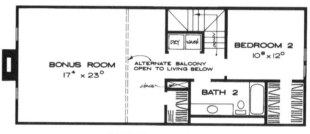

SECOND FLOOR

Old-Fashioned Charm

- This plan features:
- — Two bedrooms
- — Two full baths
- An old-fashioned, homespun flavor created by the use of lattice work, horizontal and vertical placement of wood siding, and full-length porches
- An open Living Room, Dining Room and Kitchen
- A Master Suite completes the first level
- Wood floors throughout adding a touch of country

FIRST FLOOR — 835 SQ. FT.
SECOND FLOOR — 817 SQ. FT.

TOTAL LIVING AREA:
1,652 SQ. FT.

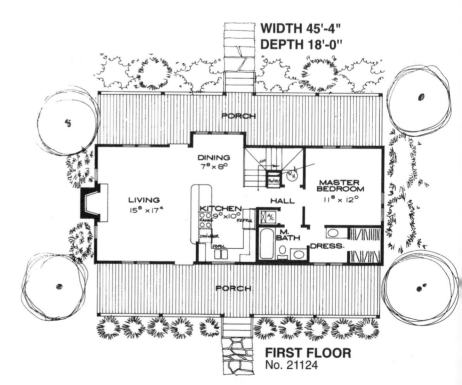

WIDTH 45'-4"
DEPTH 18'-0"

FIRST FLOOR
No. 21124

To order your Blueprints, call 1-800-235-5700

© 1995 Donald A. Gardner Architects, Inc.   B. NATHAN

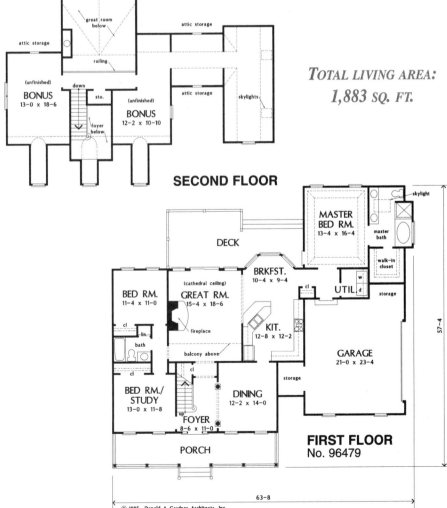

SECOND FLOOR

TOTAL LIVING AREA:
1,883 SQ. FT.

FIRST FLOOR
No. 96479

© 1995 Donald A Gardner Architects, Inc.

Growing Families Take Note

■ This plan features:
— Three bedrooms
— Two full baths

■ Unlimited options for the second floor bonus area

■ Columns accenting the dining room, adjacent to the Foyer

■ Great room, open to the Kitchen and Breakfast Room, enlarged by a cathedral ceiling

■ Living and entertaining space expands to the deck

■ Master Suite topped by a tray ceiling and includes a walk-in closet, skylit bath with garden tub and a double vanity

■ Flexible bedroom/study shares a bath with another bedroom

FIRST FLOOR — 1,803 SQ. FT.
SECOND FLOOR — 80 SQ.FT.
GARAGE & STORAGE — 569 SQ. FT.
BONUS SPACE — 918 SQ. FT.

Refer to **Pricing Schedule C** on the order form for pricing information

© 1995 Donald A. Gardner Architects, Inc.

Casually Elegant

■ This plan features:

— Three bedrooms

— Two full baths

■ Arched windows, dormers and charming front and back porches with columns creating country flavoring

■ Central Great Room topped by a cathedral ceiling, a fireplace and a clerestory window

■ Breakfast bay for casual dining is open to the Kitchen

■ Columns accenting the entryway into the formal Dining Room

■ Cathedral ceiling crowning the Master Bedroom

■ Master Bath with skylights, whirlpool tub, shower, and a double vanity

■ Two additional bedrooms sharing a bath located between the rooms

MAIN FLOOR — 1, 561 SQ. FT.
GARAGE & STORAGE — 346 SQ. FT.

TOTAL LIVING AREA:
1,561 SQ. FT.

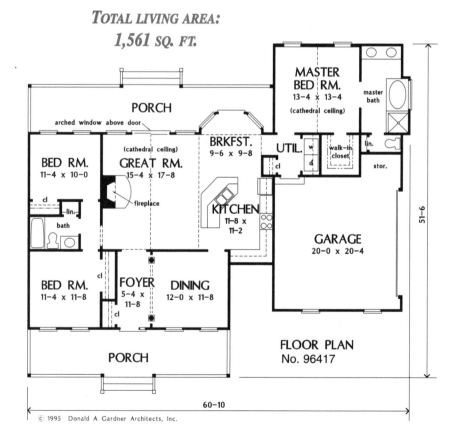

FLOOR PLAN
No. 96417

© 1995 Donald A Gardner Architects, Inc.

To order your Blueprints, call 1-800-235-5700

Refer to **Pricing Schedule A** on the order form for pricing information

Quaint Starter Home

■ This plan features:

— Three bedrooms

— Two full baths

■ A vaulted ceiling giving an airy feeling to the Dining and Living Rooms

■ A streamlined Kitchen with a comfortable work area, a double sink and ample cabinet space

■ A cozy fireplace in the Living Room

■ A Master Suite with a large closet, French doors leading to the patio and a private bath

■ Two additional bedrooms sharing a full bath

■ No materials list available for this plan

MAIN AREA — 1,050 SQ. FT.

TOTAL LIVING AREA:
1,050 SQ. FT.

Floor Plan

36

42

PATIO

MASTER BEDROOM
11 X 12

BEDROOM
9 X 12

W D

KITCHEN
9 X 11

BEDROOM
9 X 10

GARAGE
12 X 24

VAULT

VAULT

DINING
9 x 10

LIVING
14 x 14

MAIN AREA
No. 92400

Refer to **Pricing Schedule E** on the order form for pricing information

One Floor Convenience

■ This plan features:

— Four bedrooms

— Three full baths

■ A distinguished brick exterior adds curb appeal

■ Formal Entry/Gallery opens to large Living Room with hearth fireplace set between windows overlooking Patio and rear yard

■ Efficient Kitchen with angled counters and serving bar easily serves Breakfast Room, Patio and formal Dining Room

■ Corner Master Bedroom enhanced by a vaulted ceiling and pampering bath with a large walk-in closet

■ Three additional bedrooms with walk-in closets have access to full baths

■ No materials list available

MAIN FLOOR —2,675 SQ. FT.
GARAGE — 638 SQ. FT.

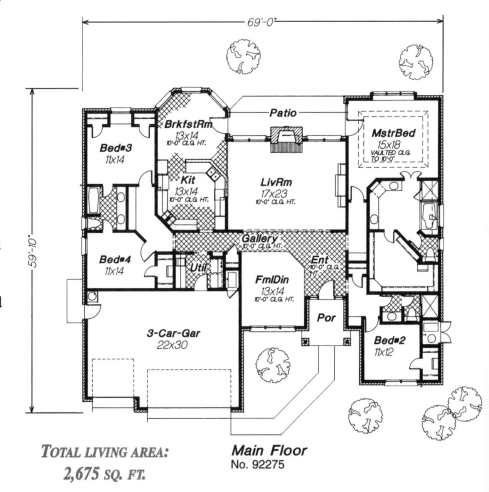

TOTAL LIVING AREA:
2,675 SQ. FT.

Main Floor
No. 92275

An
EXCLUSIVE DESIGN
By Patrick Morabito, A.I.A. Architect

FIRST FLOOR — 1,947 SQ. FT.
SECOND FLOOR — 1,390 SQ. FT.
BASEMENT — 1,947 SQ. FT.
GARAGE — 680 SQ. FT.
DECK — 322 SQ. FT.

WIDTH 86'-8"
DEPTH 49'-4"

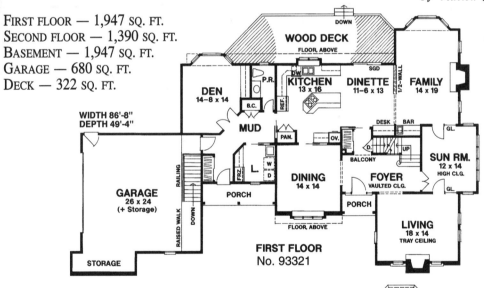

FIRST FLOOR
No. 93321

TOTAL LIVING AREA:
3,337 SQ. FT.

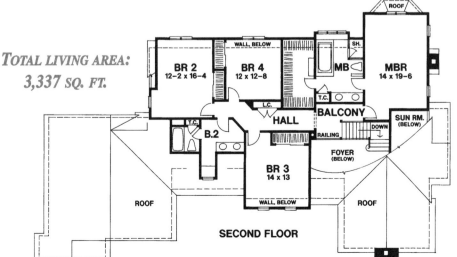

SECOND FLOOR

A *Tasteful Elegance*

■ This plan features:

— Four bedrooms

— Two full and one half baths

■ A Foyer with a vaulted ceiling, giving a great first impression

■ A Kitchen with a cooktop island and many built-in amenities

■ A Dinette with sliding glass doors to a wooden deck

■ A large Family Room with a beamed ceiling, bay window and a cozy fireplace

■ A tray ceiling as the crowning touch to the formal Living Room, which also has a terrific fireplace

■ A Master Bedroom with a stepped ceiling, double vanity private bath and a huge walk-in closet

■ Three additional bedrooms, with ample closet space, share use of a full hall Bath

■ No materials list is available for this plan

Refer to **Pricing Schedule D** on the order form for pricing information

Soft Arches Accent Country Design

■ This plan features:

— Four or five bedrooms

— Two full and one half baths

■ Entry Porch with double dormers and doors

■ Pillared arches frame Foyer, Dining Room and Great Room

■ Open Great Room with optional built-ins and sliding glass doors to Verandah

■ Compact Kitchen has walk-in pantry and a counter/snackbar

■ Comfortable Master Suite with his-n-her closets and vanities

■ Corner Study/Bedroom with Lanai access has multiple uses

■ Three second floor bedrooms share a computer loft and full bath

■ No materials list available

■ Please specify a basement or slab foundation when ordering

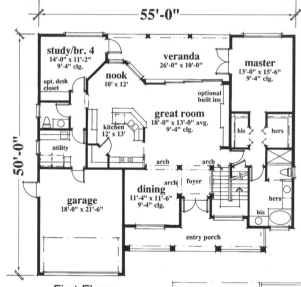

First Floor
No. 94233

FIRST FLOOR — 1,676 SQ. FT.
SECOND FLOOR — 851 SQ. FT.
GARAGE — 304 SQ. FT.

TOTAL LIVING AREA:
2,527 SQ. FT.

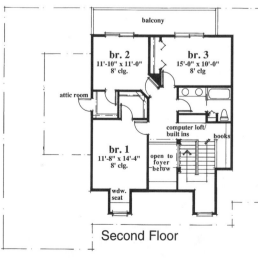

Second Floor

Refer to **Pricing Schedule A** on the order form for pricing information

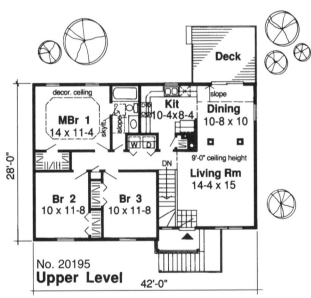

Deck

decor. ceiling

MBr 1
14 x 11-4

skyft.

slope

Kit
10-4 x 8-4

Dining
10-8 x 10

W D

9'-0" ceiling height

DN

Living Rm
14-4 x 15

Br 2
10 x 11-8

Br 3
10 x 11-8

28'-0"

No. 20195
Upper Level   42'-0"

An
EXCLUSIVE DESIGN
By Karl Kreeger

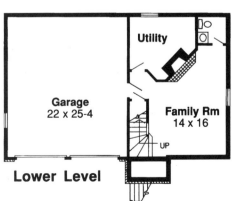

Utility

Garage
22 x 25-4

Family Rm
14 x 16

UP

Lower Level

Cozy and Restful

■ This plan features:

— Three bedrooms

— One full and one half baths

■ A decorative ceiling in the Master Bedroom with private access to the full hall sky-lit bath

■ A convenient laundry center near the bedrooms

■ An efficient Kitchen with ample counter and cabinet space and a double sink under a window

■ A Dining/Living Room combination allows easy entertaining

■ A Family Room with a cozy fireplace and convenient half bath

UPPER LEVEL — 1,139 SQ. FT.
LOWER LEVEL — 288 SQ. FT.
GARAGE — 598 SQ. FT.

TOTAL LIVING AREA:
1,427 SQ. FT.

Refer to **Pricing Schedule C** on the order form for pricing information

Country Exterior With Formal Interior

■ This plan features:

— Three bedrooms

— Two full and one half baths

■ Wrap-around Porch leads into central Foyer and formal Living and Dining rooms

■ Large Family Room with a cozy fireplace and Deck access

■ Convenient Kitchen opens to Breakfast area with a bay window and built-in pantry

■ Corner Master Bedroom with walk-in closet and appealing bath

■ Two additional bedrooms plus a bonus room share a full bath and Laundry

■ Optional basement or crawl space foundation — please specify when ordering

FIRST FLOOR — 1,046 SQ. FT.
SECOND FLOOR — 1,022 SQ. FT.
BONUS — 232 SQ. FT.

SECOND FLOOR PLAN

TOTAL LIVING AREA:
2,068 SQ. FT.

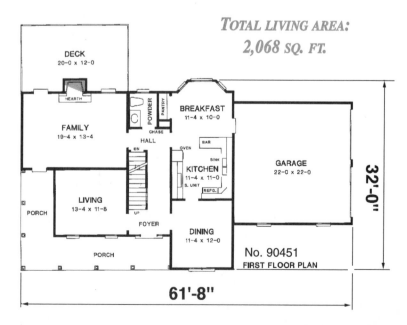

No. 90451
FIRST FLOOR PLAN

61'-8"

32'-0"

To order your Blueprints, call 1-800-235-5700

Contemporary with Cozy Front Porch

This plan features:

— Four bedrooms

— Two full and one half baths

■ A welcoming front porch

■ A Foyer that opens to a balcony above, giving a first impression of spaciousness

■ A Living Room and Dining Room that flow into each other, allowing for ease in entertaining

■ An efficient, well-appointed Kitchen that is equipped with a peninsula counter that doubles as an eating bar

■ A Breakfast Area that has easy access to a wood deck and a view of the fireplace in the Family Room

■ A Master Suite that includes a pan ceiling, private Master Bath and walk-in closet

■ Three additional bedrooms that share a full hall bath

■ No materials list is available for this plan

FIRST FLOOR — 1,028 SQ. FT.
SECOND FLOOR — 1,013 SQ. FT.
BASEMENT — 1,019 SQ. FT.
GARAGE — 479 SQ. FT.

TOTAL LIVING AREA:
2,041 SQ. FT.

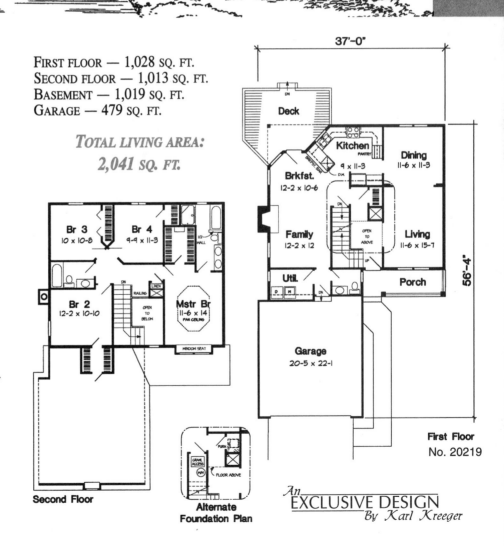

Br 3 10 x 10-8

Br 4 9-9 x 11-3

1/2 HALL

Br 2 12-2 x 10-10

LINEN

RAILING

OPEN TO BELOW

Mstr Br 11-6 x 14 PAN CEILING

WINDOW SEAT

Second Floor

CRAWL ACCESS

FLOOR ABOVE

PURR

Alternate Foundation Plan

37'-0"

DN

Deck

Kitchen PANTRY

Dining 11-6 x 11-3

Brkfst. 12-2 x 10-6

9 x 11-3

DW

OPEN TO ABOVE

Family 12-2 x 12

OPEN TO ABOVE

Living 11-6 x 15-7

UP

Util.

DN

Porch

Garage 20-5 x 22-1

56'-4"

First Floor
No. 20219

An EXCLUSIVE DESIGN *By Karl Kreeger*

Refer to **Pricing Schedule C** on the order form for pricing information

Farm Style with Dormers

- This plan features:
- — Three bedrooms
- — Two full and one half baths
- A vaulted ceiling topping the entry area
- An open layout between the Family Room, Nook and Kitchen areas
- A large work island with a built-in range top, and a walk-in pantry highlighting the unique corner angled Kitchen
- A large bay window allowing streaming natural light to brighten the Nook area
- Family Room with a wood stove and an adjacent door leading to the patio
- A vaulted ceiling crowning the Master Suite, including a walk-in closet, an oversized tub and a second vanity outside the water closet

- Two additional bedrooms on the lower level share a full bath
- An Activity Room keeping noise on the lower level

FIRST FLOOR — 1,370 SQ. FT.
LOWER LEVEL — 810 SQ. FT.
GARAGE — 564 SQ. FT.

TOTAL LIVING AREA:
2,180 SQ. FT.

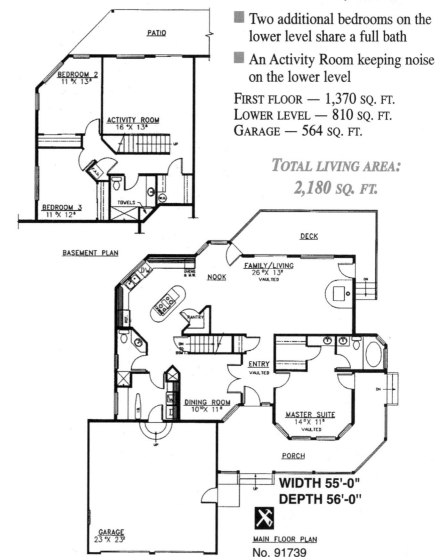

WIDTH 55'-0"
DEPTH 56'-0"

MAIN FLOOR PLAN
No. 91739

To order your Blueprints, call 1-800-235-5700

© 1995 Donald A. Gardner Architects, Inc.

S. NATHAN.

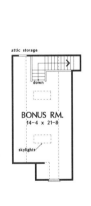

attic storage

BONUS RM.
14-4 x 21-8

skylights

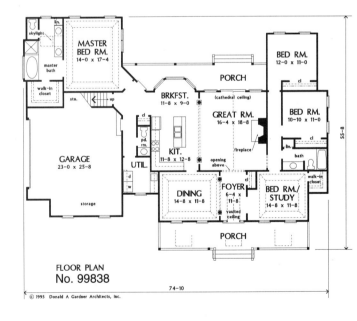

skylight

lin.

MASTER
BED RM.
14-0 x 17-4

master
bath

walk-in
closet

sto.

up

GARAGE
23-0 x 25-8

storage

UTIL.

d
w

BRKFST.
11-8 x 9-0

cl.

pd.
rm.

KIT.
11-8 x 12-8

opening
above

DINING
14-8 x 11-8

FOYER
6-4 x
11-8

vaulted
ceiling

PORCH

(cathedral ceiling)

GREAT RM.
16-4 x 18-8

fireplace

cl.

BED RM./
STUDY
14-8 x 11-8

BED RM.
12-0 x 11-0

cl.

BED RM.
10-10 x 11-0

lin.

bath

cl.

walk-in
closet

55-8

PORCH

FLOOR PLAN
No. 99838

74-10

© 1995 Donald A Gardner Architects, Inc.

TOTAL LIVING AREA:
2,192 SQ. FT.

Designed for Today's Family

■ This plan features:

— Three bedrooms

— Two full baths

■ Volume and nine foot ceilings add elegance to a comfortable, open layout

■ Secluded bedrooms designed for pleasant retreats

■ Airy Foyer topped by a vaulted dormer allowing natural light to streaming in

■ Formal Dining Room delineated from the Foyer by columns topped with a tray ceiling

■ Extra flexibility in the front bedroom as it could double as a study

■ Tray ceiling, skylights and a garden tub, in the bath highlight the Master Suite

MAIN FLOOR — 2,192 SQ. FT.
GARAGE & STORAGE — 582 SQ. FT.

Refer to **Pricing Schedule C** on the order form for pricing information

Distinctive Design

■ This plan features:

— Three bedrooms

— Two full and one half baths

■ Living Room is distinguished by warmth of bayed window and French doors leading to Family Room

■ Built-in curio cabinet adds interest to formal Dining Room

■ Well-appointed Kitchen with island cooktop and Breakfast area designed to save you steps

■ Family Room with fireplace for informal gatherings

■ Spacious Master Bedroom suite with vaulted ceiling over decorative window and plush dressing area with double walk-in closet, dual vanity and a corner whirlpool tub

■ Secondary bedrooms share a double vanity bath

SECOND FLOOR

FIRST FLOOR — 1,093 SQ. FT.
SECOND FLOOR — 905 SQ. FT.
BASEMENT — 1,093 SQ. FT.
GARAGE — 527 SQ. FT.

TOTAL LIVING AREA:
1,998 SQ. FT.

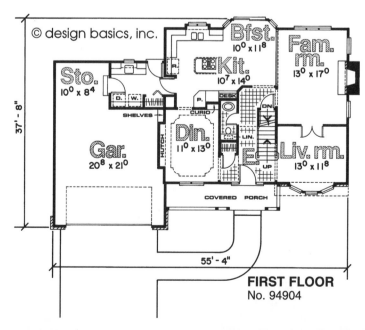

FIRST FLOOR
No. 94904

172

To order your Blueprints, call 1-800-235-5700

©1992 Donald A. Gardner Architects, Inc.

B. NATHAN

SECOND FLOOR PLAN

SITTING
10-0 x 3-4

BED RM.
12-8 x 11-0

MASTER
BED RM.
17-0 x 14-6

bath

walk-in
closet

cl

lin.

cl

down

BED RM.
12-8 x 11-6

down

BONUS RM.
12-0 x 28-2

BED RM.
12-9 x 10-0

cl

master bath

cl

Elegant Country Farmhouse

■ This plan features:

—Four bedrooms

—Two full and one half bath

■ Spacious Great Room open to the Kitchen/Breakfast area

■ Living Room/Study distinguished by elegant columns

■ Breakfast area and Great Room open to a covered porch and deck beyond

■ Great Room and Living/Study include cozy fireplaces

■ Second floor Master Suite with sitting bay, walk-in closet, whirlpool tub, shower, and dual vanity

FIRST FLOOR — 1,381 SQ. FT.
SECOND FLOOR — 1,189 SQ. FT.
GARAGE & STORAGE — 747 SQ. FT.
BONUS — 400 SQ. FT.

TOTAL LIVING AREA:
2,570 SQ. FT.

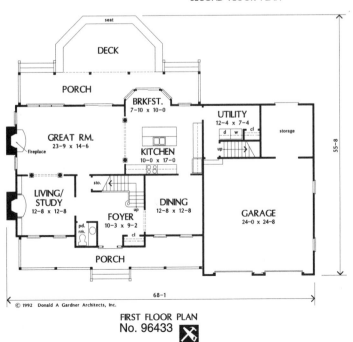

seat

DECK

PORCH

BRKFST.
7-10 x 10-0

UTILITY
12-4 x 7-4

d w cl

storage

GREAT RM.
23-9 x 14-6

KITCHEN
10-0 x 17-0

fireplace

up

sto.

LIVING/
STUDY
12-8 x 12-8

DINING
12-8 x 12-8

GARAGE
24-0 x 24-8

FOYER
10-3 x 9-2

pd.
rm.

cl

up

PORCH

55-8

68-1

© 1992 Donald A Gardner Architects, Inc.

FIRST FLOOR PLAN
No. 96433

Refer to **Pricing Schedule D** on the order form for pricing information

Quality Living

■ This plan features:

— Four bedrooms

— Two full and two half baths

■ Farmhouse front porch shelters the double door entrance

■ Sunken formal Living Room, enhanced by a focal point fireplace, a large windowed bay and a stepped ceiling

■ Elegant formal Dining Room with corners angled to form an octagon

■ A large, fully equipped Kitchen with a central island situated between formal and informal dining

■ A luxurious Master Bedroom Suite with sitting area, a walk-in closet, two linear closets and a deluxe bath

■ A Studio area above the garage

FIRST FLOOR — 1,217 SQ. FT.
SECOND FLOOR — 1,249 SQ. FT.
BASEMENT — 1,217 SQ. FT.
GARAGE — 431 SQ. FT.

TOTAL LIVING AREA:
2,466 SQ. FT.

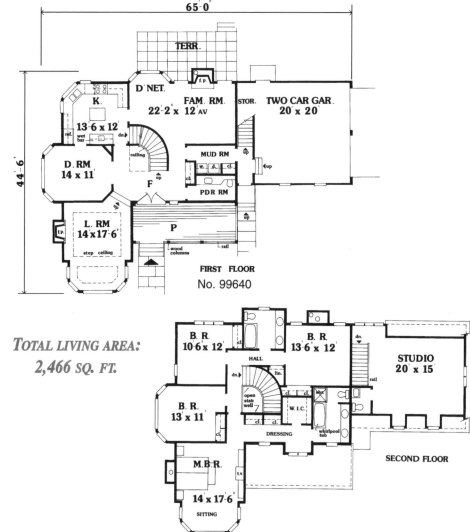

FIRST FLOOR
No. 99640

SECOND FLOOR

174

FIRST FLOOR — 2,190 SQ. FT.
SECOND FLOOR — 920 SQ. FT.
GARAGE — 624 SQ. FT.

TOTAL LIVING AREA:
3,110 SQ. FT.

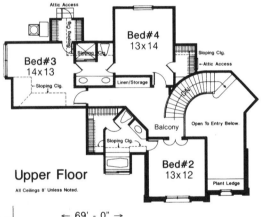

Attic Storage

Attic Access

Bed#4
13 x 14

Bed#3
14 x 13

Sloping Clg.

Sloping Clg.

Attic Access

Linen/Storage

Balcony

Open To Entry Below.

Bed#2
13 x 12

Plant Ledge

Upper Floor

All Ceilings 8' Unless Noted.

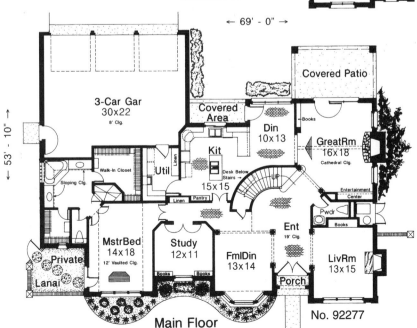

← 69' - 0" →

← 53' - 10" →

3-Car Gar
30 x 22
8' Clg.

Covered Patio

Covered Area

Din
10 x 13

Books

GreatRm
16 x 18
Cathedral Clg.

Walk-In Closet

Sloping Clg.

Util

Linen

Kit
15 x 15

Desk Below Stairs

UP

Entertainment Center

Linen

Pantry

Pwdr

Books

Ent
19' Clg.

MstrBed
14 x 18
12' Vaulted Clg.

Private

Lanai

Study
12 x 11

Books   Books

FmlDin
13 x 14

Porch

LivRm
13 x 15

No. 92277

Main Floor

Impressive Fieldstone Facade

■ This plan features:

— Four bedrooms

— Three full and one half baths

■ Double door leads into two-story entry with a curved staircase

■ Formal Living Room features a marble hearth fireplace, triple window and built-in book shelves

■ Formal Dining Room has columns and a lovely bay window

■ Efficient Kitchen offers cooktop/ work island, Utility/Garage entry and serving counter for informal dining

■ Great Room with entertainment center, fieldstone fireplace, cathedral ceiling and access to Covered Patio

■ Vaulted ceiling crowns Master Bedroom offering a plush bath and two walk-in closets

■ Three second floor bedrooms have walk-in closets

■ No materials list available

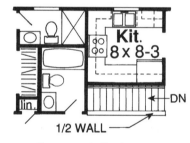

Refer to **Pricing Schedule A** on the order form for pricing information

An
EXCLUSIVE DESIGN
By Marshall Associates

Champagne Style on a Soda-Pop Budget

■ This plan features:

— Three bedrooms

— Two full baths

■ Multiple gables, circle-top windows, and a unique exterior setting this delightful Ranch apart in any neighborhood

■ Living and Dining Rooms flowing together to create a very roomy feeling

■ Sliding doors leading from the Dining Room to a covered patio

■ A Master Bedroom with a private Bath

MAIN AREA — 988 SQ. FT.
BASEMENT — 988 SQ. FT.
GARAGE — 280 SQ. FT
OPTIONAL 2-CAR GARAGE — 384 SQ. FT.

TOTAL LIVING AREA:
988 SQ. FT.

Kit.
8 x 8-3

1/2 WALL ◄—

Basement Option

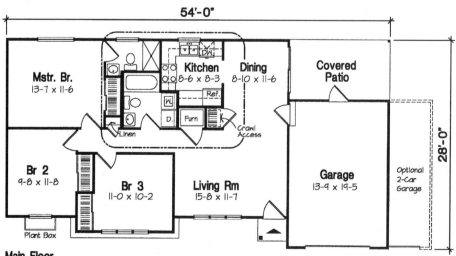

Main Floor
No. 24302

To order your Blueprints, call 1-800-235-5700

Refer to **Pricing Schedule E** on the order form for pricing information

SECOND FLOOR

FIRST FLOOR — 1,563 SQ. FT.
SECOND FLOOR — 1,315 SQ. FT.
BASEMENT — 1,547 SQ. FT.
GARAGE — 434 SQ. FT.

TOTAL LIVING AREA:
2,878 SQ. FT.

English Country Architecture

■ This plan features:

— Four bedrooms

— Four full baths

■ Open, island Kitchen blends with the Breakfast Room and directly accesses the Dining Room

■ Spacious Family Room crowned by a vaulted ceiling has a fireplace and wetbar

■ Formal Living Room is accented by a fireplace and adjoins the formal Dining Room

■ Each of the four upstairs bedrooms have access to a full bath and a generous closet

■ Master Bedroom lavishly accommodated with a private bath and an extra-large walk-in closet

■ Please specify a basement or crawl space foundation when ordering

■ No materials list available

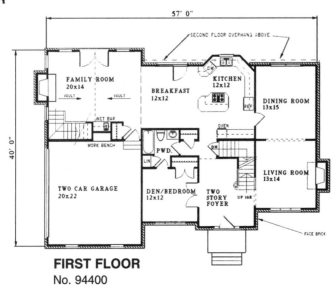

FIRST FLOOR
No. 94400

Refer to **Pricing Schedule C** on the order form for pricing information

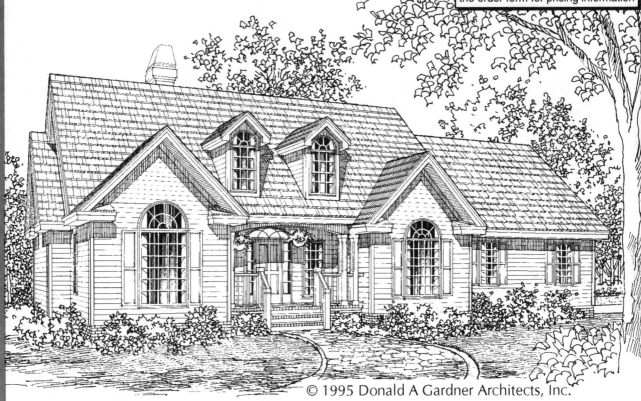

© 1995 Donald A Gardner Architects, Inc.

Great Room With Columns

■ This plan features:

— Three bedrooms

— Two full baths

■ Great Room crowned with a cathedral ceiling and accented by columns and a fireplace

■ Tray ceilings and arched picture windows accent front bedroom and the Dining Room

■ Secluded Master Suite highlighted by a tray ceiling and contains a bath with skylight, a garden tub and spacious walk-in closet

■ Two additional bedrooms share a full bath

■ Please specify a crawl space or basement foundation when ordering this plan

MAIN FLOOR — 1,879 SQ. FT.
GARAGE — 485 SQ. FT.

TOTAL LIVING AREA:
1,879 SQ. FT.

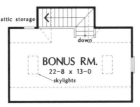

attic storage

down

BONUS RM.
22-8 x 13-0
skylights

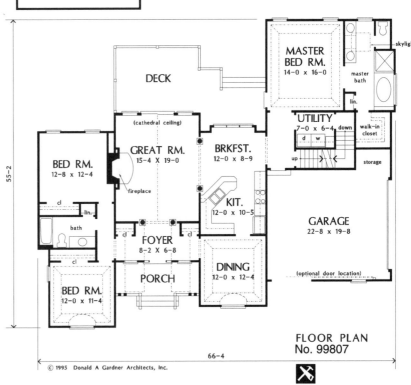

DECK

(cathedral ceiling)

GREAT RM.
15-4 x 19-0

fireplace

BED RM.
12-8 x 12-4

cl

lin.

bath

cl

cl

FOYER
8-2 X 6-8

cl

BED RM.
12-0 x 11-4

PORCH

BRKFST.
12-0 x 8-9

KIT.
12-0 x 10-5

DINING
12-0 x 12-4

MASTER BED RM.
14-0 x 16-0

skylight

master bath

lin.

UTILITY
7-0 x 6-4

d   w

down

walk-in closet

up

storage

GARAGE
22-8 x 19-8

(optional door location)

55-2

66-4

FLOOR PLAN
No. 99807

© 1995 Donald A Gardner Architects, Inc.

To order your Blueprints, call 1-800-235-5700

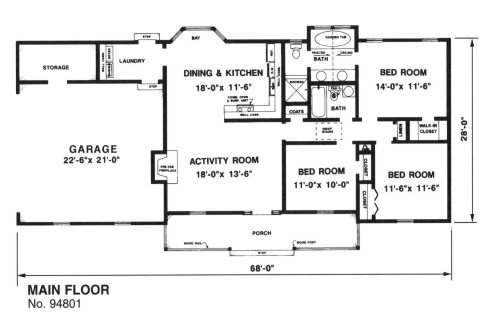

MAIN FLOOR
No. 94801

GARAGE
22'-6"x 21'-0"

STORAGE

LAUNDRY

STEP

BAY

DINING & KITCHEN
18'-0"x 11'-6"

COMB. OVEN
& SURF. UNIT

WALL CABS.

WALL CABS.

GARDEN TUB

VAULTED CEILING

BATH

SHOWER

COATS

BATH

FAN

DISAP. STAIRS

BED ROOM
14'-0"x 11'-6"

LINEN

WALK-IN CLOSET

PRE-FAB FIREPLACE

ACTIVITY ROOM
18'-0"x 13'-6"

BED ROOM
11'-0"x 10'-0"

CLOSET

CLOSET

BED ROOM
11'-6"x 11'-6"

28'-0"

PORCH

WOOD RAIL

STEP

WOOD POST

68'-0"

A Comfortable Informal Design

■ This plan features:

— Three bedrooms

— Two full baths

■ Warm, country front Porch with wood details

■ Spacious Activity Room enhanced by a pre-fab fireplace

■ Open and efficient Kitchen/Dining area highlighted by bay window, adjacent to Laundry and Garage entry

■ Corner Master Bedroom offers a pampering bath with a garden tub and double vanity topped by a vaulted ceiling

■ Two additional bedrooms with ample closets, share a full bath

■ This plan is available with a slab or crawl space foundation — please specify when ordering

MAIN FLOOR — 1,300 SQ. FT.
GARAGE — 576 SQ. FT.

TOTAL LIVING AREA:
1,300 SQ. FT.

Refer to **Pricing Schedule A** on the order form for pricing information

Affordable Style

- This plan features:

— Three bedrooms

— Two full baths

- A country porch welcomes you to an Entry hall with a convenient closet

- A well-appointed Kitchen boasts a double sink, ample counter and storage space, a peninsula eating bar and a built-in hutch

- A terrific Master Suite including a private bath and a walk-in closet

- A Dining Room that flows from the Great Room and into the Kitchen that includes sliding glass doors to the deck

- A Great Room with a cozy fireplace that can also be enjoyed from the Dining area

- Two additional bedrooms share a full hall bath

- No materials list is available for this plan

MAIN AREA
No. 91753

MAIN AREA — 1,490 SQ. FT.
COVERED PORCH — 120 SQ. FT.
BASEMENT — 1,490 SQ. FT.
GARAGE — 579 SQ. FT.
WIDTH — 58'-0"
DEPTH — 61'-0"

TOTAL LIVING AREA:
1,490 SQ. FT.

FIRST FLOOR — 905 SQ. FT.

SECOND FLOOR — 863 SQ. FT.
BASEMENT — 905 SQ. FT.
GARAGE — 487 SQ. FT.

TOTAL LIVING AREA:
1,768 SQ. FT.

© design basics, inc.

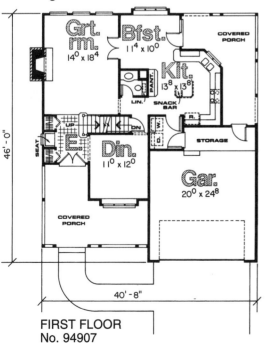

SECOND FLOOR

Mbr.
13⁰ x 14⁴
CATHEDRAL CEILING

SEAT

WHIRLPOOL

SKYLIGHT

LIN.

DN

Br. 2
10⁰ x 12⁰

Br.3
11⁰ x 10⁰

10'-0" CLG.

SEAT

FIRST FLOOR
No. 94907

Grt. rm.
14⁰ x 18⁴

Bfst.
11⁴ x 10⁰

COVERED PORCH

Kit.
13⁸ x 13⁸

PANT.

LIN.

SNACK BAR

UP

DN

SEAT

Din.
11⁰ x 12⁰

W.

D.

R.

STORAGE

Gar.
20⁰ x 24⁸

COVERED PORCH

46' - 0"

40' - 8"

Victorian Accents

- This plan features:
  — Three bedrooms
  — Two full and one half baths
- Covered Porch and double doors lead into Entry accented by a window seat and curved banister staircase
- Decorative windows overlooking the backyard and a large fireplace highlight Great Room
- A hub Kitchen with an island/snack bar and large pantry acesses the formal Dining Room and Breakfast area
- Powder room, laundry area, Garage entry and storage nearby to Kitchen
- Cathedral ceiling crowns Master Bedroom with two walk-in closets, dual vanity and a whirlpool tub
- Two additional bedrooms, one with a vaulted ceiling above a window seat, share a full bath

Refer to **Pricing Schedule E** on the order form for pricing information

Roomy and Rustic

■ This plan features:

— Four bedrooms

— Three full and one half baths

■ Cathedral Porch leads into easy-care Entry and formal Living Room with fieldstone fireplace

■ Hub Kitchen with curved peninsula serving counter convenient to Breakfast area, Covered Patio, Family Room, Utility/Garage entry and Dining Room

■ Corner Master Bedroom enhanced by vaulted ceiling, plush bath and a huge walk-in closet

■ Three additional bedrooms with walk-in closets and private access to a full bath

■ No materials list available

MAIN FLOOR — 3,079 SQ. FT.
GARAGE — 630 SQ. FT.

TOTAL LIVING AREA:
3,079 SQ. FT.

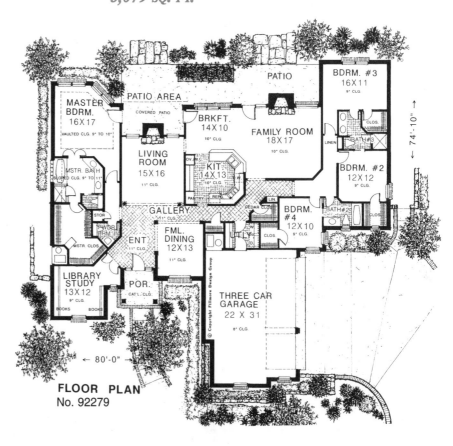

FLOOR PLAN
No. 92279

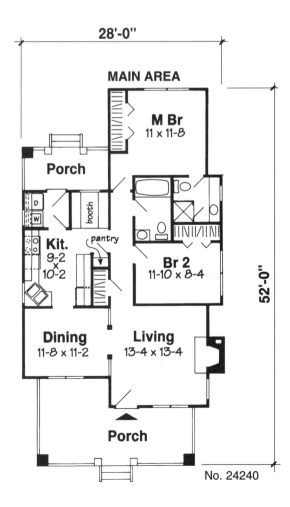

28'-0"

MAIN AREA

M Br
11 x 11-8

Porch

Kit.
9-2 x 10-2

pantry

Br 2
11-10 x 8-4

52'-0"

Dining
11-8 x 11-2

Living
13-4 x 13-4

Porch

No. 24240

Be in Tune with the Elements

■ This plan features:

— Two bedrooms

— One full baths and one three quarter bath

■ Cozy front porch to enjoy three seasons

■ A simple design allowing breezes to flow from front to back, heat to rise to the attic and cool air to settle

■ A fireplaced Living Room

■ A formal Dining Room next to the Kitchen

■ A compact Kitchen with a breakfast nook and a pantry

■ A rear entrance with a covered porch

■ A Master Suite with a private bath

MAIN AREA — 964 SQ. FT.

TOTAL LIVING AREA:
964 SQ. FT.

Refer to **Pricing Schedule B** on the order form for pricing information

Savor the Summer

■ This plan features:

— Four bedrooms

— Two full and one half baths

■ A efficient home with a friendly front Porch and a practical back porch

■ A cozy fireplace and a boxed window with a built-in seat in the Living Room

■ A formal Dining Room opening to front entrance and Kitchen

■ A well-equipped Kitchen with an old-fashion booth and ample cabinet and counter space adjoining Laundry area and back porch

■ A convenient, first floor Master Suite with two closets and a private Bath

■ Three additional bedrooms on second floor share a full hall bath

FIRST FLOOR — 931 SQ. FT.
SECOND FLOOR — 664 SQ. FT.

TOTAL LIVING AREA:
1,595 SQ. FT.

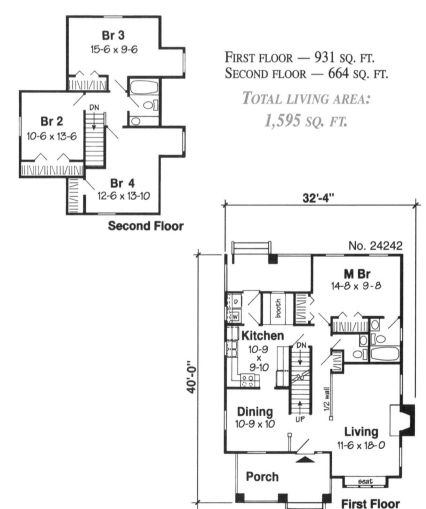

Br 3
15-6 x 9-6

Br 2
10-6 x 13-6

DN

Br 4
12-6 x 13-10

Second Floor

32'-4"

No. 24242

M Br
14-8 x 9-8

booth

Kitchen
10-9 x 9-10

D
W

DN

1/2 wall

40'-0"

Dining
10-9 x 10

UP

Living
11-6 x 18-0

Porch

seat

First Floor

To order your Blueprints, call 1-800-235-5700

An
EXCLUSIVE DESIGN
By Ahmann Design Inc.

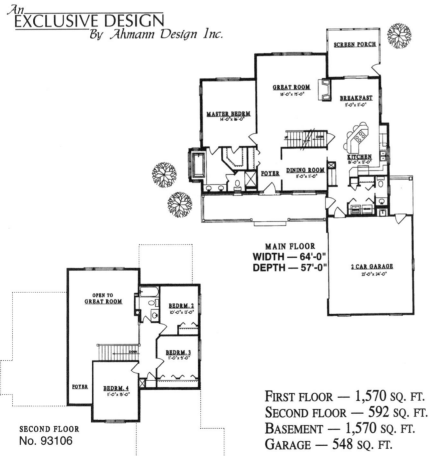

SCREEN PORCH

GREAT ROOM
18'-0" x 15'-0"

BREAKFAST
11'-0" x 11'-0"

MASTER BEDRM.
14'-0" x 16'-0"

KITCHEN
15'-0" x 12'-0"

FOYER

DINING ROOM
11'-0" x 11'-0"

MAIN FLOOR
WIDTH — 64'-0"
DEPTH — 57'-0"

2 CAR GARAGE
22'-0" x 24'-0"

OPEN TO
GREAT ROOM

BEDRM. 2
10'-0" x 12'-0"

BEDRM. 3
11'-0" x 9'-0"

FOYER

BEDRM. 4
11'-0" x 15'-0"

SECOND FLOOR
No. 93106

FIRST FLOOR — 1,570 SQ. FT.
SECOND FLOOR — 592 SQ. FT.
BASEMENT — 1,570 SQ. FT.
GARAGE — 548 SQ. FT.

TOTAL LIVING AREA:
2,162 SQ. FT.

Contemporary Plan

■ This plan features:

— Four bedrooms

— Two full and one half baths

■ A classic country front porch
leading to the Foyer and second
entrance into Mud/Utility Room

■ A formal Dining Room located
next to the Kitchen

■ A large country Kitchen with an
angled island/eating bar and a
Breakfast Area

■ A Screened Porch expanding
living space in the warmer
months

■ An expansive Great Room
enhanced by a cozy fireplace

■ A Master Suite highlighted by a
Jacuzzi, surrounded by a wall of
glass

■ Three additional bedrooms
sharing a full bath

■ No materials list available

Refer to **Pricing Schedule E** on the order form for pricing information

B. NATHAN

© 1995 Donald A. Gardner Architects, Inc.

Welcoming Exterior

■ This plan features:

— Four bedrooms

— Two full and one half baths

■ Columns between the Foyer and Living Room/Study

■ Transom windows over French doors open up the Living Room/Study to the Front Porch, while a generous Family Room accesses the covered Back Porch

■ Deluxe Master Suite is topped by a tray ceiling and includes a bath with a sunny garden tub bay and ample closet space

■ Bonus room is accessed from the second floor and ready to expand family living space for future needs

FIRST FLOOR — 1,483 SQ. FT.
SECOND FLOOR — 1,349 SQ. FT.
GARAGE — 738 SQ. FT.

TOTAL LIVING AREA:
2,832 SQ. FT.

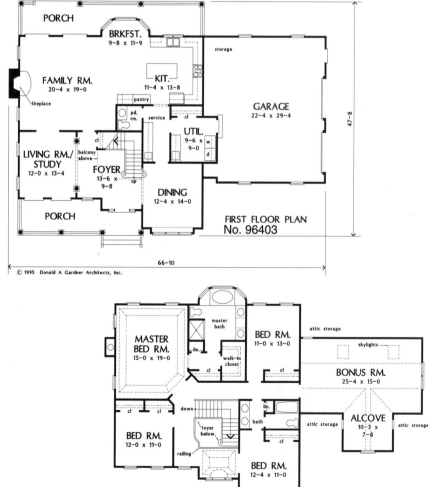

FIRST FLOOR PLAN
No. 96403

© 1995 Donald A Gardner Architects, Inc.

SECOND FLOOR PLAN

186

Refer to **Pricing Schedule C** on the order form for pricing information

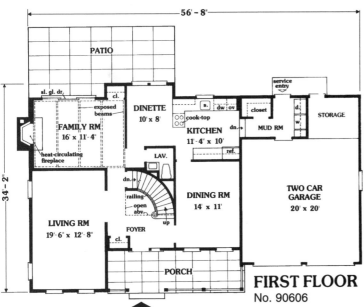

FIRST FLOOR
No. 90606

56' – 8"

34' – 2"

PATIO

sl. gl. dr.   cl.

DINETTE
10' x 8'

service entry

s.   dw   ov

closet

d.

w.

STORAGE

exposed beams

FAMILY RM
16' x 11'-4"

cook-top

KITCHEN
11'-4' x 10'

dn.

MUD RM

heat-circulating fireplace

LAV.

ref.

dn.

railing

open abv.

DINING RM
14' x 11'

TWO CAR GARAGE
20' x 20'

LIVING RM
19'-6" x 12'-8"

FOYER

up

cl.

PORCH

SECOND FLOOR

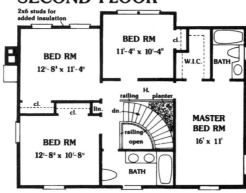

2x6 studs for added insulation

BED RM
12'-8' x 11'-4"

BED RM
11'-4" x 10'-4"

cl.

W.I.C.

BATH

cl.

cl.

lin.

railing   H.   planter

dn.

BED RM
12'-8' x 10'-8"

railing

open

MASTER BED RM
16' x 11'

BATH

Traditional Elements Combine in Friendly Colonial

■ This plan features:

— Four bedrooms

— Two full and one half baths

■ A beautiful circular stair ascending from the central Foyer and flanked by the formal Living Room and Dining Room

■ Exposed beams, wood paneling, and a brick fireplace wall in the Family Room

■ A separate Dinette opening to an efficient Kitchen

FIRST FLOOR — 1,099 SQ. FT.
SECOND FLOOR — 932 SQ. FT.
BASEMENT — 1,023 SQ. FT.

TOTAL LIVING AREA:
2,031 SQ. FT.

Refer to **Pricing Schedule C** on the order form for pricing information

Classic Exterior with Modern Interior

- This plan features:

— Three or four bedrooms

— Two full and one half baths

- Front Porch leads into an open Foyer and Great Room beyond, which is accented by a sloped ceiling, corner fireplace and multiple windows

- An efficient Kitchen with a cooktop island, walk-in pantry, a bright Dining Area and nearby Screened Porch, Laundry and Garage entry

- Deluxe Master Bedroom wing with a decorative ceiling, large walk-in closet and plush bath

- Three or four bedrooms on the second floor share a double vanity bath

- No materials list available

FIRST FLOOR — 1,348 SQ. FT.
SECOND FLOOR — 528 SQ. FT.

TOTAL LIVING AREA:
1,876 SQ. FT.

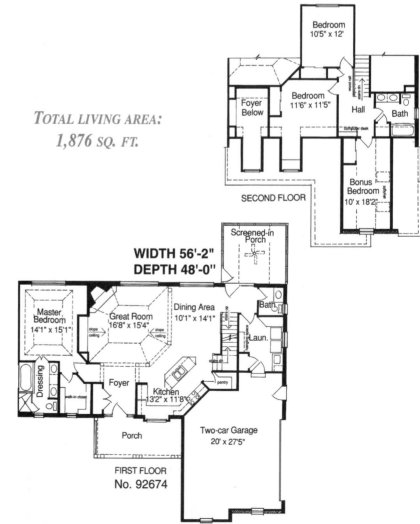

SECOND FLOOR

Bedroom 10'5" x 12'

Bedroom 11'6" x 11'5'

Foyer Below

Hall

Bath

Bonus Bedroom 10' x 18'2"

WIDTH 56'-2"
DEPTH 48'-0"

Screened-in Porch

Master Bedroom 14'1" x 15'1"

Great Room 16'8" x 15'4"

Dining Area 10'1" x 14'1"

Bath

Laun.

Dressing

Foyer

Kitchen 13'2" x 11'8"

pantry

walk-in closet

Porch

Two-car Garage 20' x 27'5"

FIRST FLOOR
No. 92674

© 1997 Donald A Gardner Architects, Inc.

B. NATHAN

PLAN NO. 99801

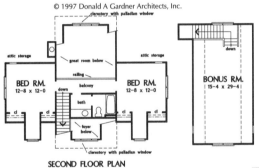

© 1997 Donald A Gardner Architects, Inc.

clerestory with palladian window

attic storage

great room below

attic storage

railing

BED RM.
12-8 x 12-0

balcony

down

BED RM.
12-8 x 12-0

cl.

cl.

bath

cl.

cl.

foyer below

clerestory with palladian window

SECOND FLOOR PLAN

BONUS RM.
15-4 x 29-4

down

TOTAL LIVING AREA:
2,188 SQ. FT.

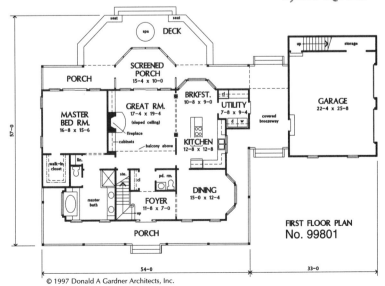

seat   DECK   seat
spa

SCREENED
PORCH
15-4 x 10-0

PORCH

up   storage

GARAGE
22-4 x 25-8

BRKFST.
10-8 x 9-0

UTILITY
7-8 x 9-4

cl.

MASTER
BED RM.
16-8 x 15-6

GREAT RM.
17-4 x 19-4
(sloped ceiling)
fireplace

cabinets

balcony above

KITCHEN
12-8 x 12-8

d w

covered breezeway

walk-in
closet

lin.

sto.

cl.

pd. rm.

master
bath

FOYER
11-8 x 7-0

up

DINING
15-0 x 12-4

57-0

PORCH

FIRST FLOOR PLAN
No. 99801

54-0   33-0

© 1997 Donald A Gardner Architects, Inc.

Columns Punctuate the Interior Space

■ This plan features:

— Three bedrooms

— Two full and one half baths

■ A two-story Great Room and Foyer, both with dormer windows, welcome natural light into this graceful country classic with a wrap-around porch

■ Large Kitchen, featuring a center cooking island with counter and large Breakfast area, opens to the Great Room for easy entertaining

■ Columns punctuate the interior spaces and a separate Dining Room provides a formal touch to the plan

■ Master Bedroom Suite, privately situated on the first floor, has a double vanity, garden tub, and separate shower

FIRST FLOOR — 1,618 SQ. FT.
SECOND FLOOR — 570 SQ. FT.
BONUS ROOM — 495 SQ. FT.
GARAGE & STORAGE — 649 SQ. FT.

Refer to **Pricing Schedule C** on the order form for pricing information

Convenient and Efficient Ranch

■ This plan features:

— Three bedrooms

— Two full and one half baths

■ A barrel vault ceiling in the Foyer

■ A stepped ceiling in both the Dinette and the formal Dining Room

■ An expansive Gathering Room with a large focal point fireplace and access to the wood deck

■ An efficient Kitchen that includes a work island and a built-in pantry

■ A luxurious Master Suite with a private bath that includes a separate tub and step-in shower

■ Two additional bedrooms that share a full hall bath

■ No materials list is available for this plan

MAIN AREA — 1,810 SQ. FT.

GARAGE — 528 SQ. FT.

TOTAL LIVING AREA:
1,810 SQ. FT.

An EXCLUSIVE DESIGN
By Patrick Morabito, A.I.A. Architect

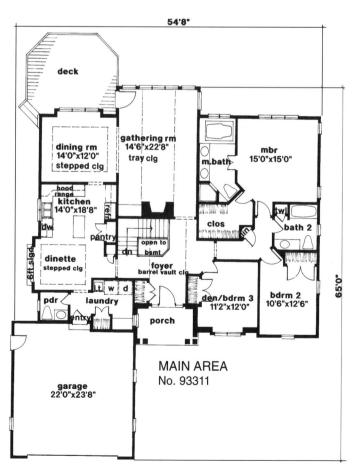

MAIN AREA
No. 93311

To order your Blueprints, call 1-800-235-5700

P L A N N O . 2 4 3 2 6

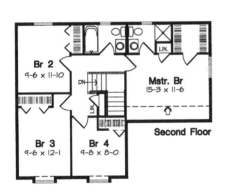

Br 2
9-6 x 11-10

Mstr. Br
15-3 x 11-6

Br 3
9-6 x 12-1

Br 4
9-8 x 8-0

DN

LIN.

Second Floor

CRAWL ACCESS

PANTRY

Crawl/Slab Option

TOTAL LIVING AREA:
1,505 SQ. FT.

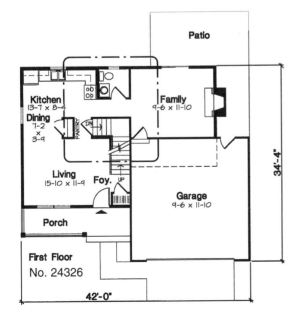

Patio

Kitchen
13-7 x 8-4

Dining
7-2 x 3-9

Family
9-6 x 11-10

PANTRY

DN

Living
15-10 x 11-9

Foy. UP

Garage
9-6 x 11-10

Porch

34'-4"

First Floor
No. 24326

42'-0"

Fireplace-Equipped Family Room

■ This plan features:

— Four bedrooms

— Two full baths and one half bath

■ A lovely front porch shading the entrance

■ A spacious Living Room that opens into the Dining Area which flows into the efficient Kitchen

■ A Family Room equipped with a cozy fireplace and sliding glass doors to a patio

■ A Master Suite with a large walk-in closet and a private bath with a step-in shower

■ Three additional bedrooms that share a full hall bath

FIRST FLOOR — 692 SQ. FT.
SECOND FLOOR — 813 SQ. FT.
BASEMENT — 699 SQ. FT.
GARAGE — 484 SQ. FT.

An
EXCLUSIVE DESIGN
By Marshall Associates

Refer to **Pricing Schedule D** on the order form for pricing information

© 1994 Donald A. Gardner Architects, Inc.

Elegance And A Relaxed Lifestyle

■ This plan features:

— Four bedrooms

— Three full baths

■ Open two-level Foyer has a palladian window which visually ties the formal Dining area to the expansive Great Room

■ Bay windows in Master Bedroom Suite and Breakfast area provide natural light, while nine foot ceilings create volume

■ Master Bedroom suite features a whirlpool tub, separate shower and his-n-her vanities

FIRST FLOOR — 1,841 SQ. FT.
SECOND FLOOR — 594 SQ. FT.
BONUS ROOM — 411 SQ. FT.
GARAGE & STORAGE — 596 SQ. FT.

TOTAL LIVING AREA:
2,435 SQ. FT.

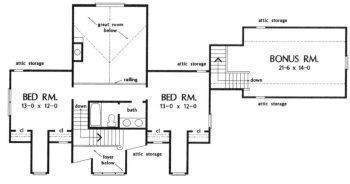

SECOND FLOOR PLAN

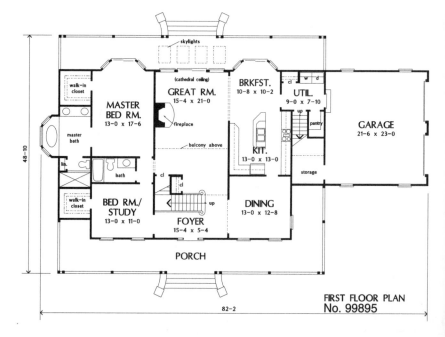

FIRST FLOOR PLAN
No. 99895

To order your Blueprints, call 1-800-235-5700

Refer to **Pricing Schedule F** on the order form for pricing information

©1993 Donald A. Gardner Architects, Inc.

PLAN NO. 96441

FIRST FLOOR — 2,357 SQ. FT.
SECOND FLOOR — 995 SQ. FT.
GARAGE & STORAGE — 975 SQ. FT.
BONUS ROOM — 545 SQ. FT.

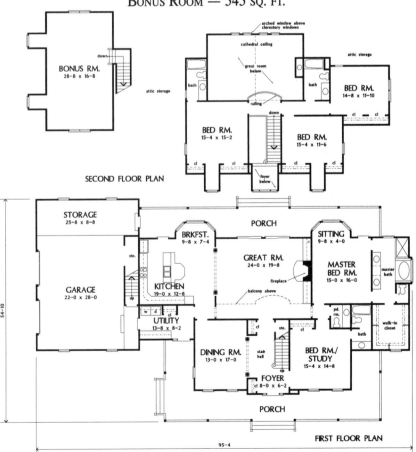

No. 96441

Classic Sprawling Farmhouse

- This plan features:
- —Five bedrooms
- —Four full and one half baths
- Two level Foyer with a palladian window, a large Great Room with a cathedral ceiling and curved balcony make a dramatic statement
- Master Suite with a bayed sitting area, plush bath and large walk-in closet
- Bedroom/study with a private full bath offers a double Master Suite option
- Master bedroom, Kitchen and Great Room access covered back porch
- Bonus room for future expansion
- Three additional bedrooms on the second floor, one with a private bath

TOTAL LIVING AREA:
3,352 SQ. FT.

To order your Blueprints, call 1-800-235-5700

Refer to **Pricing Schedule C** on the order form for pricing information

© 1997 Donald A. Gardner Architects, Inc.

B. NATHAN

Outdoor Living Options

- This plan features:
- —Three bedrooms
- —Two full baths
- Living areas that are open and casual
- Great Room crowned by a cathedral ceiling which continues out to the screened porch
- Kitchen opens to a sunny breakfast bay and is adjacent to the formal Dining Room
- Master suite topped by a tray ceiling and enhanced by an indulgent bath with a roomy walk-in closet
- Two additional bedrooms sharing a full bath

MAIN FLOOR —1,609 SQ. FT.
GARAGE & STORAGE — 500 SQ. FT.

TOTAL LIVING AREA:
1,609 SQ. FT.

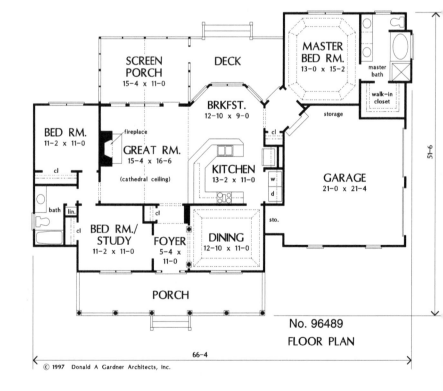

SCREEN PORCH
15-4 x 11-0

DECK

MASTER BED RM.
13-0 x 15-2

master bath

walk-in closet

BED RM.
11-2 x 11-0

fireplace

BRKFST.
12-10 x 9-0

storage

cl

GREAT RM.
15-4 x 16-6

KITCHEN
13-2 x 11-0

(cathedral ceiling)

cl

w
d

GARAGE
21-0 x 21-4

cl

bath

lin.

lin

BED RM./
STUDY
11-2 x 11-0

FOYER
5-4 x
11-0

cl

DINING
12-10 x 11-0

sto.

PORCH

51-6

No. 96489
FLOOR PLAN

66-4

© 1997 Donald A Gardner Architects, Inc.

To order your Blueprints, call 1-800-235-5700

MAIN AREA — 2,072 SQ. FT.
GARAGE — 585 SQ. FT.

WIDTH 60'-0"
DEPTH 70'-0"

TOTAL LIVING AREA:
2,072 SQ. FT.

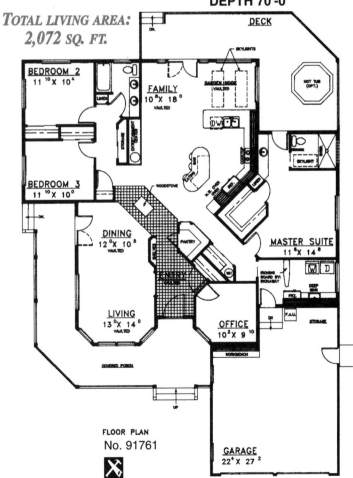

FLOOR PLAN
No. 91761

Lots of Room to Entertain

■ This plan features:

— Three bedrooms

— Two full baths

■ An open floor plan that is spacious and easy to adapt for wheelchair accessibility

■ A Kitchen, with an eating bar, that flows into the Family Room; allowing for continued conversation between the two rooms

■ Direct access to the wood deck from the Family Room that features a vaulted ceiling with skylights

■ A Master Suite enhanced by a private bath with a skylight and direct access to the wood deck

■ A combination Living Room/Dining Room making entertaining easier and more enjoyable

■ Two additional bedrooms that share a full hall bath

Refer to **Pricing Schedule E** on the order form for pricing information

Spectacular Stucco and Stone

◼ This plan features:

— Three bedrooms

— Two full and one half baths

◼ Two-story glass entry opens to Great Room with fireplace between bookcases

◼ Convenient Dining Room opens to Great Room and adjoins Kitchen

◼ Open Kitchen and Morning Room with pantry, cooktop island, fireplace and skylights, provides efficiency and relaxation

◼ Secluded Master Bedroom offers lovely bay window, two walk-in closets and a luxurious bath

◼ Two second floor bedrooms with walk-in closets and vanities, have private access to a full bath

◼ No materials list available

◼ Please specify a basement or crawl space foundation when ordering

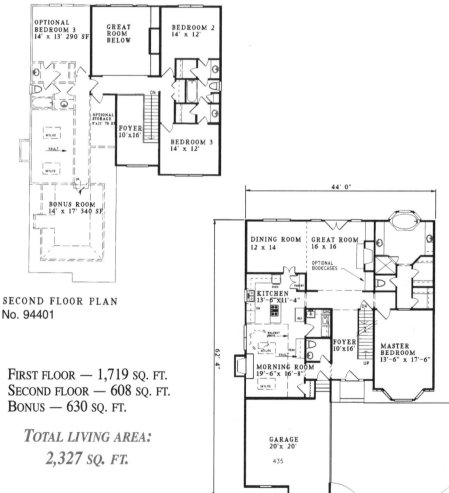

SECOND FLOOR PLAN
No. 94401

FIRST FLOOR — 1,719 SQ. FT.
SECOND FLOOR — 608 SQ. FT.
BONUS — 630 SQ. FT.

TOTAL LIVING AREA:
2,327 SQ. FT.

FIRST FLOOR PLAN

To order your Blueprints, call 1-800-235-5700

Cathedral Ceiling in Living Room and Master Suite

■ This plan features:

— Three bedrooms

— Two full baths

■ A spacious Living Room with a cathedral ceiling and elegant fireplace

■ A Dining Room that adjoins both the Living Room and the Kitchen

■ An efficient Kitchen, with double sinks, ample cabinet space and peninsula counter that doubles as an eating bar

■ A convenient hallway laundry center

■ A Master Suite with a cathedral ceiling and a private Master Bath

MAIN AREA — 1,346 SQ. FT.
GARAGE — 449 SQ. FT.

TOTAL LIVING AREA:
1,346 SQ. FT.

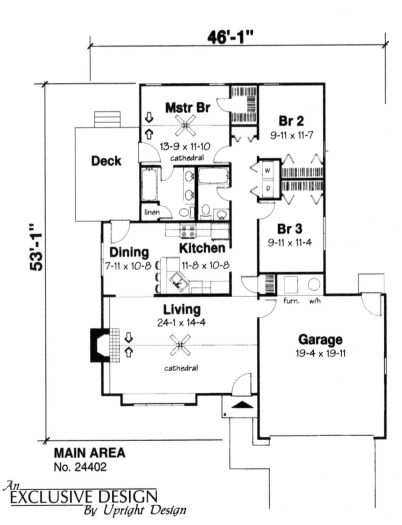

46'-1"

53'-1"

Mstr Br
13-9 x 11-10
cathedral

Deck

Br 2
9-11 x 11-7

linen

Br 3
9-11 x 11-4

Dining
7-11 x 10-8

Kitchen
11-8 x 10-8

Living
24-1 x 14-4
cathedral

fum.   w/h

Garage
19-4 x 19-11

MAIN AREA
No. 24402

An EXCLUSIVE DESIGN
By Upright Design

Refer to **Pricing Schedule C** on the order form for pricing information

Built In Entertainment Center for Family Fun

■ This plan features:

— Four bedrooms

— Two full and one half baths

■ A heat-circulating fireplace in the Living Room framed by decorative pilasters that support dropped beams

■ A convenient mudroom providing access to the two-car Garage

■ A spacious Master Suite with a separate dressing area

FIRST FLOOR — 1,094 SQ. FT.
SECOND FLOOR — 936 SQ. FT.
GARAGE — 441 SQ. FT.

TOTAL LIVING AREA:
2,030 SQ. FT.

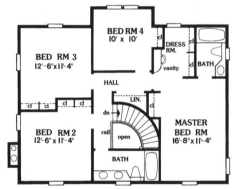

SECOND FLOOR PLAN

BED RM 4
10' x 10'
DRESS RM.
BATH
vanity
BED RM 3
12'-6"x11'-4"
HALL
cl cl cl cl
dn
LIN.
cl
BED RM 2
12'-6" x 11'-4"
rail   open
MASTER BED RM
16'-8"x11'-4"
BATH

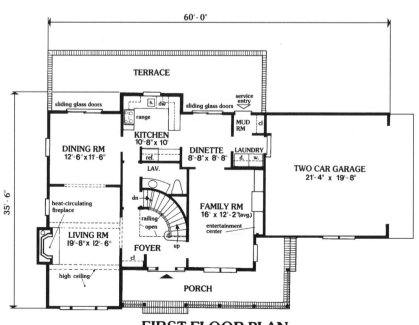

60'-0"

TERRACE

sliding glass doors
s.   dw
range
sliding glass doors
service entry
MUD RM   cl

KITCHEN
10'-8" x 10'
DINING RM
12'-6" x 11'-6"
ref.
DINETTE
8'-8" x 8'-8"
LAUNDRY
d.   w.

LAV.
dn
heat-circulating fireplace
railing
open
LIVING RM
19'-8" x 12'-6"
FOYER
up
FAMILY RM
16' x 12'-2" (avg.)
entertainment center
TWO CAR GARAGE
21'-4" x 19'-8"

35'-6"

cl
high ceiling

PORCH

FIRST FLOOR PLAN
No. 90615

© 1995 Donald A Gardner Architects, Inc.

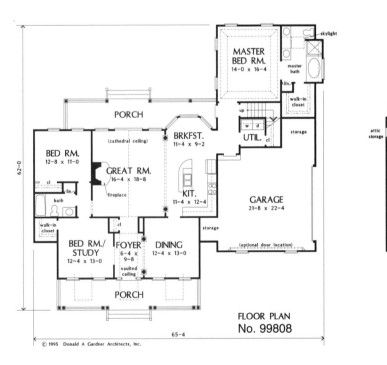

PORCH

MASTER BED RM.
14-0 x 16-4

skylight

master bath

lin.

walk-in closet

up

storage

BED RM.
12-8 x 11-0

(cathedral ceiling)

GREAT RM.
16-4 x 18-8

fireplace

BRKFST.
11-4 x 9-2

KIT.
11-4 x 12-4

cl

w d

UTIL.

cl

GARAGE
21-8 x 22-4

attic storage

storage

down

BONUS RM.
12-8 x 22-4

skylights

cl

bath

walk-in closet

BED RM./ STUDY
12-4 x 13-0

FOYER
6-4 x 9-8

vaulted ceiling

DINING
12-4 x 13-0

storage

(optional door location)

PORCH

62-0

65-4

FLOOR PLAN
No. 99808

© 1995 Donald A Gardner Architects, Inc.

TOTAL LIVING AREA:
1,832 SQ. FT.

Classic Country Farmhouse

◼ This plan features:

— Three bedrooms

— Two full baths

◼ Dormers, arched windows and multiple columns give this home country charm

◼ Foyer, expanded by vaulted ceiling, accesses Dining Room, Bedroom/Study and Great Room

◼ Expansive Great Room, with hearth fireplace opens to rear Porch and efficient Kitchen

◼ Tray ceiling adds volume to private Master Bedroom with plush bath and walk-in closet

◼ Extra room for growth offered by Bonus Room with skylight

MAIN FLOOR — 1,832 SQ. FT.
BONUS ROOM — 425 SQ. FT.
GARAGE & STORAGE — 562 SQ. FT.

Refer to **Pricing Schedule D** on the order form for pricing information

© 1993 Donald A. Gardner Architects, Inc.

Covered Porches Front and Back

■ This plan features:

— Three bedrooms

— Two full and one half baths

■ Open floor plan plus a Bonus Room great for today's family needs

■ Two-story Foyer with palladian, clerestory window and balcony overlooking Great Room

■ Great Room with cozy fireplace provides perfect gathering place

■ Columns visually separate Great Room from Breakfast area and smart, U-shaped Kitchen

■ Privately located Master Bedroom accesses Porch and luxurious Master Bath with separate shower and double vanity

FIRST FLOOR — 1,632 SQ. FT.
SECOND FLOOR — 669 SQ. FT.
BONUS ROOM — 528 SQ. FT.
GARAGE & STORAGE — 707 SQ. FT.

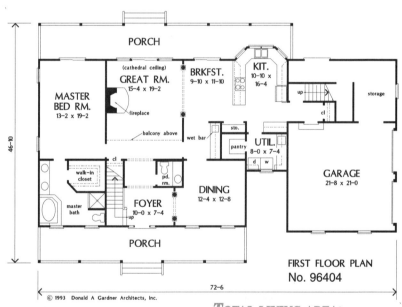

FIRST FLOOR PLAN
No. 96404

© 1993 Donald A Gardner Architects, Inc.

TOTAL LIVING AREA:
2,301 SQ. FT.

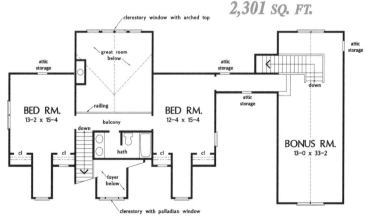

SECOND FLOOR PLAN

To order your Blueprints, call 1-800-235-5700

Refer to **Pricing Schedule F** on the order form for pricing information

An
EXCLUSIVE DESIGN
By Patrick Morabito, A.I.A. Architect

SECOND FLOOR

FIRST FLOOR — 1,880 SQ. FT.
SECOND FLOOR — 1,465 SQ. FT.
BASEMENT — 1,880 SQ. FT.
GARAGE — 900 SQ. FT.

TOTAL LIVING AREA:
3,345 SQ. FT.

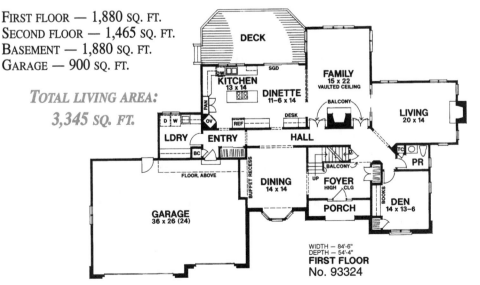

WIDTH — 84'-6"
DEPTH — 54'-4"
FIRST FLOOR
No. 93324

Traditional Classic

This plan features:

— Four bedrooms

— Two full and one half baths

A distinguished bay window and buffet recess, enhancing the formal Dining Room

A splendid fireplace accenting the formal Living Room that includes French doors between it and the Family Room

A vaulted ceiling and a second fireplace in the expansive Family Room

A well-appointed Kitchen equipped with an island and Dinette Area with sliding glass doors to the wooden deck

A special tray ceiling adding a decorative touch to the second floor Master Suite

Three additional bedrooms that share a full hall bath

No materials list is available for this plan

PLAN NO. 24246

Refer to **Pricing Schedule D** on the order form for pricing information

For the Executive

■ This plan features:

— Three bedrooms

— Three full baths

■ A sloped, two-story ceiling in the Foyer giving a definite impression of distinguished elegance

■ A double door entrance into the formal Living Room adding anticipation while the corner fireplace adds warmth and atmosphere to the room

■ A built-in china alcove in the formal Dining Room

■ A U-shaped Kitchen directly accessed form the Dining Room

■ A built-in pantry, breakfast bar, double sink and ample counter and storage space in the Kitchen adding to its efficiency

■ A second corner fireplace accenting the Family Room

■ A whirlpool corner tub, two vanities, a separate shower and a walk-in closet in the Master Suite

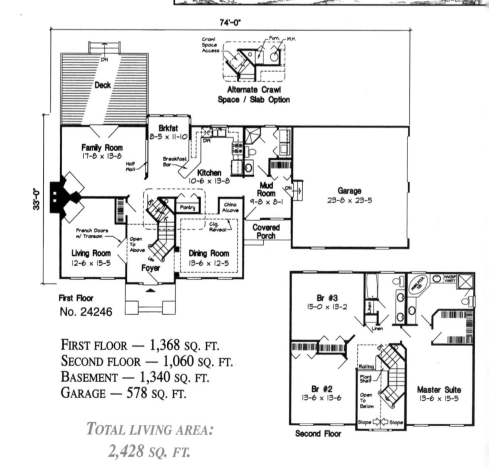

FIRST FLOOR — 1,368 SQ. FT.
SECOND FLOOR — 1,060 SQ. FT.
BASEMENT — 1,340 SQ. FT.
GARAGE — 578 SQ. FT.

TOTAL LIVING AREA:
2,428 SQ. FT.

Refer to **Pricing Schedule C** on the order form for pricing information

TOTAL LIVING AREA:
2,192 SQ. FT.

An
EXCLUSIVE DESIGN
By Jannis Vann & Associates, Inc.

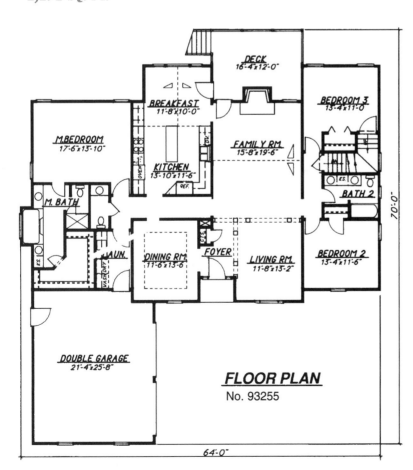

FLOOR PLAN
No. 93255

DECK
16'-4"x12'-0"

BREAKFAST
11'-8"x10'-0"

BEDROOM 3
13'-4"x11'-0"

FAMILY RM.
15'-8"x19'-6"

M.BEDROOM
17'-6"x13'-10"

KITCHEN
13'-10"x11'-6"

BATH 2

M. BATH

LAUN.

DINING RM.
11'-6"x13'-6"

FOYER

LIVING RM.
11'-8"x13'-2"

BEDROOM 2
13'-4"x11'-6"

DOUBLE GARAGE
21'-4"x25'-8"

70'-0"

64'-0"

Varied Roof Lines Add Interest

▪ This plan features:

— Three bedrooms

— Two full and one half baths

▪ A modern, convenient floor plan

▪ Formal areas located at the front of the home

▪ A decorative ceiling in the Dining Room

▪ Columns accent the Living Room

▪ A large Family Room with a cozy fireplace and direct access to the deck

▪ An efficient Kitchen located between the formal Dining Room and the informal Breakfast Room

▪ A private Master Suite with a Master Bath and walk-in closet

▪ Two additional bedrooms share a full hall bath

MAIN AREA — 2,192 SQ. FT.
BASEMENT — 2,192 SQ. FT.
GARAGE — 564 SQ. FT.

Refer to **Pricing Schedule B** on the order form for pricing information

Open Plan is Full of Air & Light

■ This plan features:

—Three bedrooms

—Two full and one half baths

■ Foyer open to the Family Room and highlighted by a fireplace

■ Dining Room, with a sliding glass door to rear yard, adjoins Family Room

■ Kitchen and Nook in an efficient open layout

■ Second floor Master Suite topped by tray ceiling over the bedroom and a vaulted ceiling over the lavish bath

■ No materials list is available for this plan

■ When ordering this plan—please specify a basement or crawl space foundation

FIRST FLOOR — 767 SQ. FT.
SECOND FLOOR — 738 SQ. FT.
BONUS ROOM — 240 SQ. FT.
BASEMENT — 767 SQ. FT.

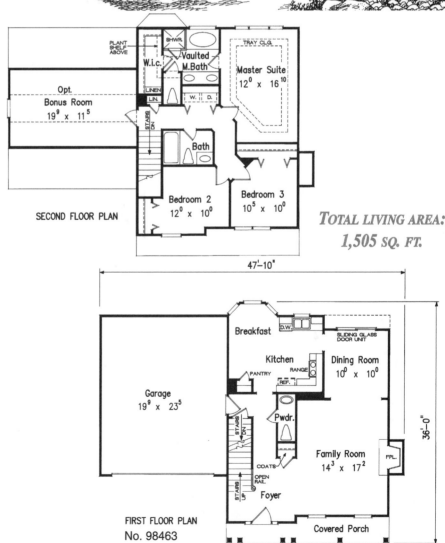

PLANT SHELF ABOVE

SHWR.

W.i.c.

Vaulted M.Bath

TRAY CLG.

Master Suite
12⁰ x 16¹⁰

Opt.
Bonus Room
19⁹ x 11⁵

LINEN
LIN.

W. D.

STAIRS DN

Bath

SECOND FLOOR PLAN

Bedroom 2
12⁰ x 10⁰

Bedroom 3
10⁵ x 10⁰

TOTAL LIVING AREA:
1,505 SQ. FT.

47'-10"

Breakfast

D.W.

SLIDING GLASS DOOR UNIT

Kitchen

RANGE

Dining Room
10⁰ x 10⁰

PANTRY

REF.

Garage
19⁹ x 23⁵

STAIRS DN

Pwdr.

36'-0"

Family Room
14³ x 17²

FPL.

COATS

OPEN RAIL

STAIRS UP

Foyer

FIRST FLOOR PLAN
No. 98463

Covered Porch

To order your Blueprints, call 1-800-235-5700

Refer to **Pricing Schedule B** on the order form for pricing information

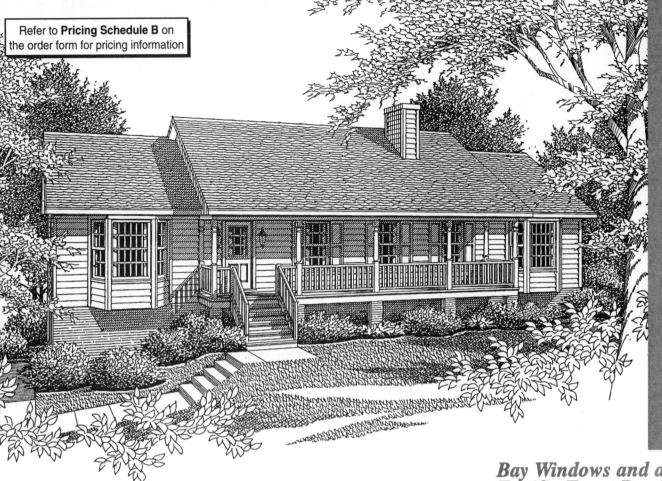

TOTAL LIVING AREA:
1,778 SQ. FT.

An
EXCLUSIVE DESIGN
By Jannis Vann & Associates, Inc.

SUNDECK
16'. 0" X 14'. 0"

MAIN AREA
No. 93261

DINING RM.
12'. 6" X 11'. 6"

KIT.
9'. 0" X11'. 4"

BREAKFAST
9'8" X 13'. 6"

PANT.

BEDROOM 3
13'. 6" X 11'-0"

REF

DESK

48'-0"

M.BEDROOM
13'. 6" X17'. 2"

FOYER
5'8" X11'.6"

LIVING AREA
19'8" X 15'.6"

BEDROOM 2
13'. 6" X 11'. 8"

PORCH
34'.0"X6'.0"

ASSOC., INC.

62'-0"

Bay Windows and a Terrific Front Porch

■ This plan features:

— Three bedrooms

— Two full baths

■ A Country style front porch

■ An expansive Living Area that includes a fireplace

■ A Master Suite with a private Master Bath and a walk-in closet, as well as a bay window view of the front yard

■ An efficient Kitchen that serves the sunny Breakfast Area and the Dining Room with equal ease

■ A built-in pantry and a desk add to the conveniences in the Breakfast Area

■ Two additional bedrooms that share the full hall bath

■ A convenient main floor Laundry Room

MAIN AREA — 1,778 SQ. FT.
BASEMENT — 1,008 SQ. FT.
GARAGE — 728 SQ. FT.

Refer to **Pricing Schedule B** on the order form for pricing information

An
EXCLUSIVE DESIGN
By Jannis Vann & Associates, Inc.

Old Fashioned Country Porch

■ This plan features:

— Three bedrooms

— Two full and one half baths

■ A Traditional front Porch, with matching dormers above and a garage hidden below, leading into a open, contemporary layout

■ A Living Area with a cozy fireplace visible from the Dining Room for warm entertaining

■ A U-shaped, efficient Kitchen featuring a corner, double sink and pass-thru to the Dining Room

■ A convenient half bath with a laundry center on the first floor

■ A spacious, first floor Master Suite with a lavish Bath including a double vanity, walk-in closet and an oval, corner window tub

■ Two large bedrooms with dormer windows, on the second floor, sharing a full hall bath

■ No materials list available

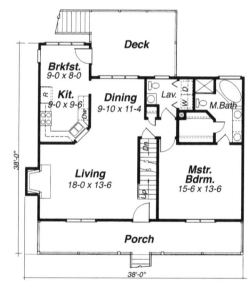

Deck

Brkfst.
9-0 x 8-0

Kit.
9-0 x 9-6

Dining
9-10 x 11-4

Lav.

M.Bath

Living
18-0 x 13-6

Mstr. Bdrm.
15-6 x 13-6

Porch

38'-0"

38'-0"

FIRST FLOOR
No. 93219

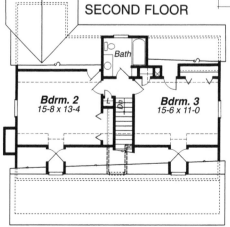

SECOND FLOOR

Bath

Bdrm. 2
15-8 x 13-4

Bdrm. 3
15-6 x 11-0

FIRST FLOOR — 1,057 SQ. FT.
SECOND FLOOR — 611 SQ. FT.
BASEMENT — 511 SQ. FT.
GARAGE — 546 SQ. FT.

TOTAL LIVING AREA:
1,668 SQ. FT.

To order your Blueprints, call 1-800-235-5700

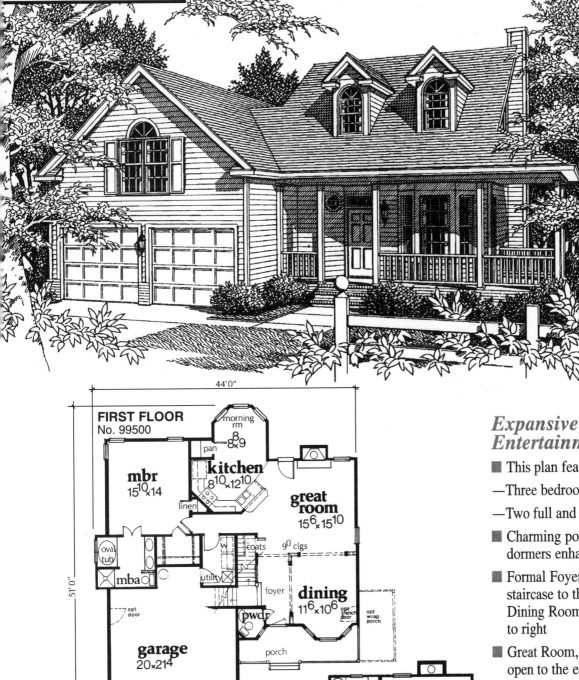

PLAN NO. 99500

FIRST FLOOR
No. 99500

44'0"

51'0"

morning rm
8x9

pan

kitchen
8^{10}×12^{10}

linen

mbr
15^{10}×14

great room
15^6×15^{10}

9^0 clgs

oval tub

mba

coats

w

utility

foyer

dining
11^6×10^6

opt french door

opt wrap porch

9^0 clgs

opt door

pwdr

garage
20×21^4

porch

An ___
EXCLUSIVE DESIGN
By Georgia Toney Cesley,
Residential Designer

First Floor — 1,218 sq. ft.
Second Floor — 864 sq. ft.
Garage — 472 sq. ft.

Total living area:
2,082 sq. ft.

ba 2
lin

br 2
11^8×15

storage

plant shelf

br 3
16×12

dormers

9' ceiling

entertainment room
16^8×20

Expansive Entertainment Room

■ This plan features:

— Three bedrooms

— Two full and one half baths

■ Charming porch and quaint dormers enhance curb appeal

■ Formal Foyer with half bath and staircase to the left, elegant Dining Room with a bay window to right

■ Great Room, with fireplace is open to the efficient Kitchen

■ First floor Master Suite has a five-piece bath and a walk-in closet

■ Entertainment room on second floor keeps playful children happy and the noise upstairs

■ No materials list is available for this plan

■ Please specify a slab or a crawl space foundation when ordering this plan

Refer to **Pricing Schedule E** on the order form for pricing information

© 1993 Donald A. Gardner Architects, Inc.

B. NATHAN

Country Porches Front and Back

■ This plan features:

—Four bedrooms

—Three full baths

■ Elegantly bayed Dining Room directly accessing the cooktop island Kitchen

■ Breakfast Bay flowing from the efficient Kitchen

■ Great Room, directly accessing the back porch, highlighted by a cathedral ceiling, a fireplace and a balcony above

■ Master Suite enhanced by a lavish private bath and a walk-in closet

■ Two additional bedrooms on the second floor sharing the full double vanity bath in the hall

FIRST FLOOR — 1,871 SQ. FT.
SECOND FLOOR — 731 SQ. FT.
BONUS ROOM — 402 SQ. FT.
GARAGE & STORAGE — 600 SQ. FT.

TOTAL LIVING AREA:
2,602 SQ. FT.

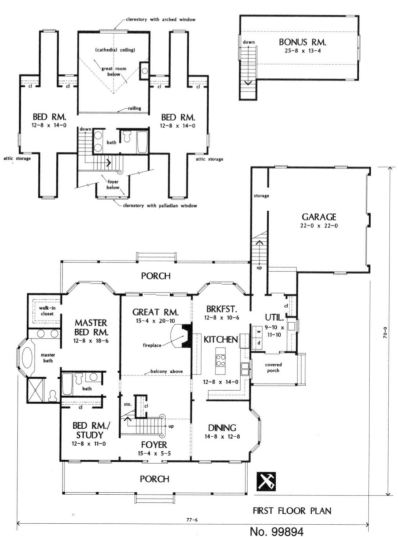

FIRST FLOOR PLAN

No. 99894

To order your Blueprints, call 1-800-235-5700

PLAN NO. 98730

MAIN AREA — 2,596 SQ. FT.
GARAGE — 1,037 SQ. FT.

TOTAL LIVING AREA:
2,596 SQ. FT.

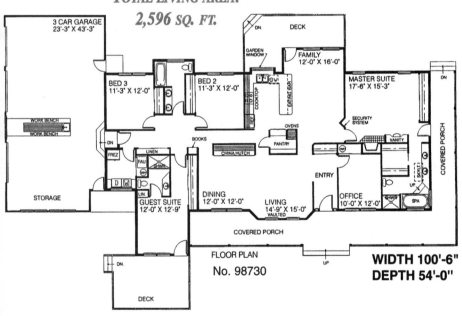

FLOOR PLAN
No. 98730

WIDTH 100'-6"
DEPTH 54'-0"

Plenty of Space for the Active Family

■ This plan features:

— Four bedrooms

— Three full baths

■ A bright Living Room with a vaulted ceiling

■ A country Kitchen stocked with built-in ovens, pantry, cooktop, garden window and an eating bar

■ A built-in hutch that also serves as a buffet in the Dining Room

■ A spacious family room leads to the rear deck

■ Lots of built-in storage space

■ A Guest Suite with a private bath

■ A luxurious Master Suite with a fireplace, vanity, spa, skylights, twin vanities and a conveniently located security system

■ Two secondary bedrooms sharing use of the twin basin, hall bath

■ No materials list is available for this plan

© 1996 Donald A. Gardner Architects, Inc.

B. NATHAN

Dramatic Dormers

■ This plan features:

—Three bedrooms

—Two full baths

■ A Foyer open to the dramatic dormer, defined by columns

■ A Dining Room augmented by a tray ceiling

■ A Great Room expands into the Kitchen and Breakfast Room

■ A privately located Master Suite, topped by a tray ceiling and pampered by a garden tub with a picture window as the focal point of the master bath

■ Two additional bedrooms, located at the opposite side share a full bath and linen closet

MAIN FLOOR — 1,685 SQ. FT.

GARAGE & STORAGE — 536 SQ. FT.

TOTAL LIVING AREA:
1,685 SQ. FT.

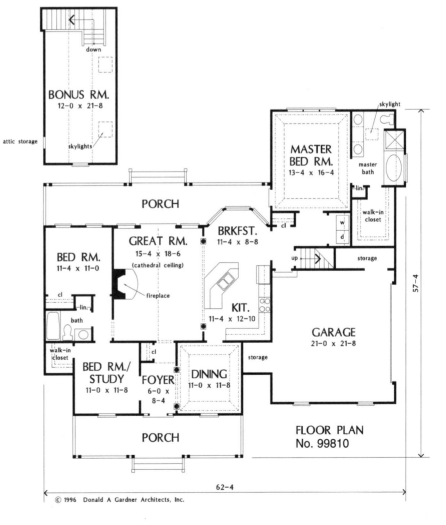

BONUS RM.
12–0 x 21–8

down

attic storage

skylights

PORCH

BED RM.
11–4 x 11–0

cl

lin.

bath

walk-in closet

BED RM./
STUDY
11–0 x 11–8

GREAT RM.
15–4 x 18–6
(cathedral ceiling)

fireplace

cl

FOYER
6–0 x
8–4

DINING
11–0 x 11–8

BRKFST.
11–4 x 8–8

KIT.
11–4 x 12–10

storage

PORCH

MASTER
BED RM.
13–4 x 16–4

skylight

master bath

lin.

walk-in closet

w
d

up

storage

GARAGE
21–0 x 21–8

cl

57–4

62–4

FLOOR PLAN
No. 99810

© 1996 Donald A Gardner Architects, Inc.

MAIN FLOOR — 1,641 SQ. FT.
GARAGE — 427 SQ. FT.

TOTAL LIVING AREA:
1,641 SQ. FT.

TERR.
76'-4"

M.B.R.
16'-8" x 12'

B.R.
10' x 13'-2"

L.R.
high ceiling
skylight
13' x 20'-6"

KIT.   DIN.
18'-6" x 10'

dw
ref.
w d
up
dn.

w.i.cl.
lin.
cl.
cl.

whirlpool tub
shower
cl.

B.R.
11' x 10'
cath. ceiling

H.
cl.

DEN
10' x 10'
cath. ceiling

F.

D.R.
11'-4" x 10'

cl.

2 - CAR GAR.
20' x 20'

PORCH

35'-10"

up

see alternate plans

H.

B.R.
10' x 10'

ALT. PLAN

H.

OFFICE
10' x 10'

ALT. PLAN

optional door

FLOOR PLAN
No. 99657

Flexible Floor Plan

■ This plan features:

— Three bedrooms

— Two full baths

■ Sheltering entrance Porch leads into an open Foyer, Living and Dining rooms

■ Comfortable Dining Room with an elegant bow window

■ Spacious Living Room enhanced by skylight, inviting fireplace, and Terrace access

■ Country-size Kitchen offers a bright Dining area with Terrace access, laundry closet and Garage entry

■ Corner Master Bedroom offers a large walk-in closet and deluxe bath with a whirlpool tub

■ Two additional bedrooms share a double vanity bath

■ Convenient Den can be easily converted to an Office or fourth bedroom

Refer to **Pricing Schedule C** on the order form for pricing information

© 1993 Donald A. Gardner Architects, Inc.

B. NATHAN

Quaint and Cozy

■ This plan features:

— Three bedrooms

— Two full and one half bath

■ Spacious floor plan with large Great Room crowned by cathedral ceiling

■ Central kitchen with angled counter opens to the breakfast area and Great Room for easy entertaining

■ Privately located Master Bedroom has a cathedral ceiling and nearby access to the deck with an optional spa

■ Operable skylights over the tub accent the luxurious Master Bath

■ Bonus room over the garage makes expanding easy

■ Please specify crawl space or basement foundation when ordering

MAIN FLOOR — 1,864 SQ. FT.
GARAGE — 614 SQ. FT.

TOTAL LIVING AREA:
1,864 SQ. FT.

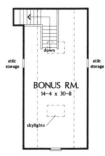

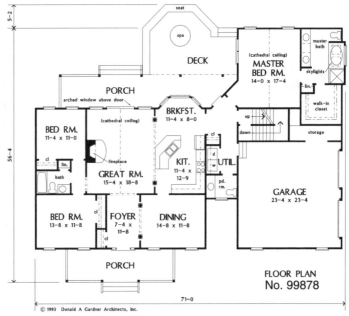

FLOOR PLAN
No. 99878

© 1993 Donald A Gardner Architects, Inc.

©1997 Donald A. Gardner Architects, Inc.

optional bedroom wall location

optional bedroom wall location

attic storage

great room below

railing

attic storage

BED RM.
12-8 x 12-4

BED RM.
12-8 x 12-4

down

sto.

lin.

bath

(cathedral ceiling)

cl

cl

cl

cl

SECOND FLOOR PLAN

BONUS RM.
21-0 x 19-3

down

attic storage

TOTAL LIVING AREA:
2,596 SQ. FT.

Wrapping Front Porch and Gabled Dormers

■ This plan features:

— Four bedrooms

— Three full baths

■ Generous Great Room with a fireplace, cathedral ceiling, and a balcony above

■ Flexible bedroom/study with a walk-in closet and an adjacent full bath

■ First floor Master Suite with a sunny bay window and a private bath with cathedral ceiling, his-n-her vanities, and a separate tub and shower

■ Bonus room over the garage for future expansion

First floor — 1,939 sq. ft.

Second floor — 657 sq. ft.

Garage & Storage — 526 sq. ft.

Bonus room — 386 sq. ft.

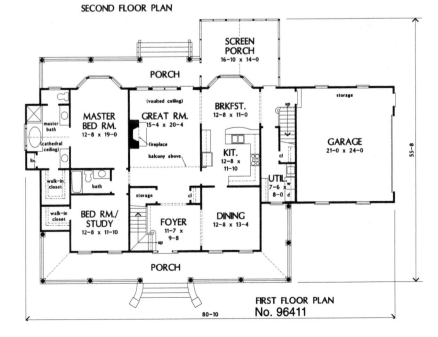

SCREEN PORCH
16-10 x 14-0

PORCH

storage

master bath

(cathedral ceiling)

MASTER BED RM.
12-8 x 19-0

GREAT RM.
15-4 x 20-4

(vaulted ceiling)

fireplace

balcony above

BRKFST.
12-8 x 11-0

up

GARAGE
21-0 x 24-0

55-8

lin.

walk-in closet

bath

KIT.
12-8 x 11-10

cl

storage

UTIL.
7-6 x 8-0

w

walk-in closet

BED RM./ STUDY
12-8 x 11-10

cl

FOYER
11-7 x 9-8

DINING
12-8 x 13-4

up

PORCH

80-10

FIRST FLOOR PLAN
No. 96411

B. NATHAN

Refer to **Pricing Schedule D** on the order form for pricing information

Whimsical Two-story Farmhouse

- ■ This plan features:
  - — Four bedrooms
  - — Three full and one half baths
- ■ Double gable with palladian, clerestory window and wrap-around Porch provide country appeal
- ■ First floor enjoys nine foot ceilings throughout
- ■ Palladian windows flood two-story Foyer and Great Room with natural light
- ■ Both Master Bedroom and Great Room access covered rear Porch
- ■ One upstairs bedroom offers private bath and walk-in closet

FIRST FLOOR — 1,346 SQ. FT.
SECOND FLOOR — 836 SQ. FT.

TOTAL LIVING AREA:
2,182 SQ. FT.

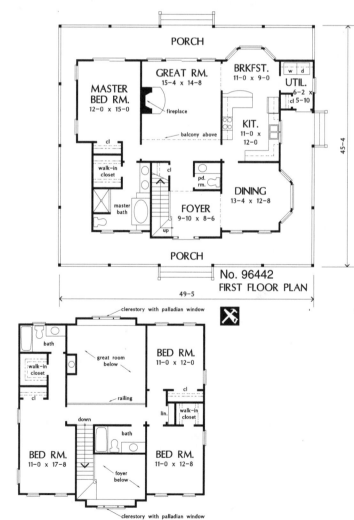

No. 96442
FIRST FLOOR PLAN

SECOND FLOOR PLAN

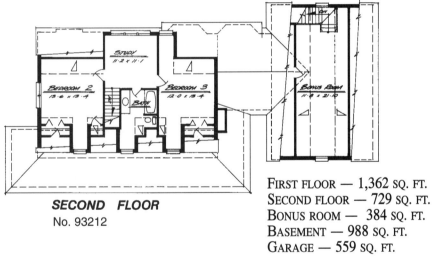

SECOND FLOOR
No. 93212

FIRST FLOOR — 1,362 SQ. FT.
SECOND FLOOR — 729 SQ. FT.
BONUS ROOM — 384 SQ. FT.
BASEMENT — 988 SQ. FT.
GARAGE — 559 SQ. FT.

An
EXCLUSIVE DESIGN
By Jannis Vann & Associates. Inc.

TOTAL LIVING AREA:
2,091 SQ. FT.

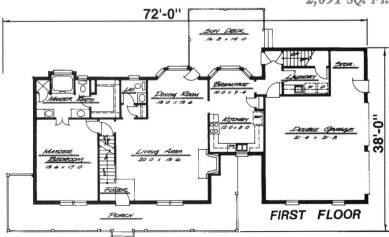

72'-0"

38'-0"

FIRST FLOOR

Old Fashioned Feel Yet Modern Comforts

■ This plan features:

— Three bedrooms

— Two full and one half baths

■ Porch and dormers give this plan an old-fashioned feel

■ Large Living Room with a cozy fireplace open to the Dining Room

■ Formal Dining Room has a bay window and direct access to the Sun Deck

■ A sunny Breakfast Nook with a bay window overlooking the deck

■ U-shaped Kitchen, efficiently arranged with ample work space and pantry

■ A first floor Master Suite has an elegant private bath

■ A second floor study or hobby room overlooking the deck and backyard

■ No materials list available

■ Please specify a basement, crawl space or slab foundation when ordering

Refer to **Pricing Schedule D** on the order form for pricing information

Watch the World Go By

■ This plan features:

— Three or four bedrooms

— Three full and one half baths

■ A large wrap-around porch

■ An immense Living Room, highlighted by a fieldstone fireplace

■ A Study or Guest Room with easy access to a full bath

■ A central work island Kitchen

■ Open floor plan between the Kitchen and the Family Room

■ A formal Dining Room with direct access to the Kitchen and views of the front porch

■ A Sewing room and a Utility Room separated by a powder room

■ A large Master Suite served by a private bath and huge walk-in closet

FIRST FLOOR — 1,463 SQ. FT.
SECOND FLOOR — 981 SQ. FT.
BASEMENT — 814 SQ. FT.

TOTAL LIVING AREA:
2,444 SQ. FT.

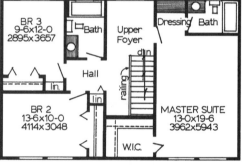

An
EXCLUSIVE DESIGN
By Westhome Planners, Ltd.

SECOND FLOOR

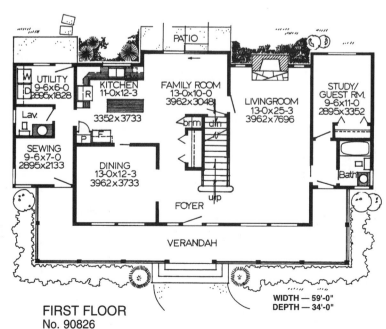

FIRST FLOOR
No. 90826

WIDTH — 59'-0"
DEPTH — 34'-0"

216 To order your Blueprints, call 1-800-235-5700

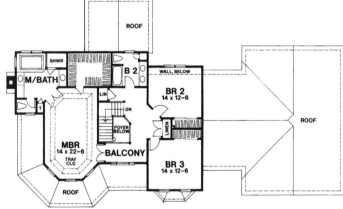

SECOND FLOOR
82-0

ROOF

M/BATH
SHWR
B 2
WALL BELOW
BR 2
14 x 12-6
LIN
DN
FOYER BELOW
MBR
14 x 22-6
TRAY CLG
BALCONY
LINEN
BR 3
14 x 12-6
ROOF
ROOF

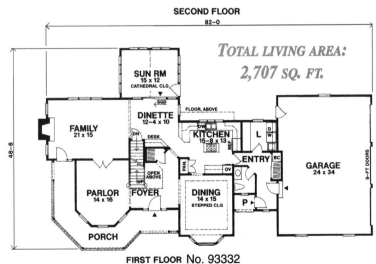

TOTAL LIVING AREA:
2,707 SQ. FT.

48-8

SUN RM
15 x 12
CATHEDRAL CLG
SGD
FLOOR ABOVE
DINETTE
12-4 x 10
DW
KITCHEN
15-8 x 13
REF
L
D
W
FAMILY
21 x 15
DN
DESK
ENTRY
BC
GARAGE
24 x 34
9-FT DOORS
OPEN ABOVE
PAN.
OV
UP
PARLOR
14 x 16
FOYER
DINING
14 x 15
STEPPED CLG
P
PORCH

FIRST FLOOR No. 93332

■ No materials list is available for this plan

FIRST FLOOR — 1,484 SQ. FT.
SECOND FLOOR — 1,223 SQ. FT.

That Old-Fashion Feeling

■ This plan features:

— Three bedrooms

— Two full and one half baths

■ A formal Parlor opening into the Family Room, with a hearth fireplace, for easy entertaining

■ A stepped ceiling accenting a charming bay window in the formal Dining Room

■ A large, island Kitchen with a double sink, a built-in pantry and a peninsula counter/eating bar leading to a large Entry with access to both the Garage and Laundry Room

■ Sliding glass doors in the Dinette leading to the Sun Room, with a cathedral ceiling

■ An elegant Master Suite with a tray ceiling, a room-sized, walk-in closet, and a plush Bath, featuring a raised, corner window tub and two vanities

An
EXCLUSIVE DESIGN
By Patrick Morabito, A.I.A. Architect

B. NATHAN

© 1993 Donald A. Gardner Architects, Inc.

Built-In Convenience

■ This plan features:

—Four bedrooms

—Three full baths

■ Formal Dining Room directly accesses efficient Kitchen for ease in serving

■ Cooktop island Kitchen with built-in pantry and ample cabinet counter space

■ Breakfast Room directly accesses the Kitchen that is defined by columns from the Great Room

■ Cathedral ceiling, cozy fireplace, and built-in cabinet highlight the Great Room as a balcony overlooks from above

■ Built-in wetbar in alcove off the Great Room

■ Plush Master Suite pampered by a private, skylit bath and a walk-in closet

FIRST FLOOR — 1783 SQ. FT.
SECOND FLOOR — 611 SQ. FT.
GARAGE — 506 SQ. FT.

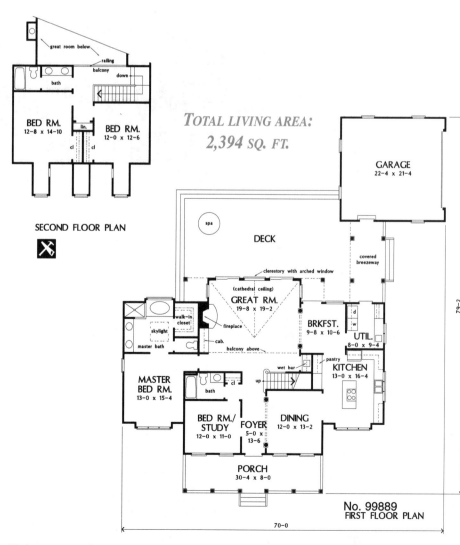

SECOND FLOOR PLAN

BED RM.
12-8 x 14-10

BED RM.
12-0 x 12-6

great room below
railing
balcony
down
bath
lin.
cl   cl

TOTAL LIVING AREA:
2,394 SQ. FT.

GARAGE
22-4 x 21-4

DECK
spa

covered breezeway

clerestory with arched window

(cathedral ceiling)
GREAT RM.
19-8 x 19-2

BRKFST.
9-8 x 10-6

UTIL.
8-0 x 9-4

walk-in closet
skylight
master bath
fireplace
cab.
balcony above

wet bar
pantry
KITCHEN
13-0 x 16-4

MASTER BED RM.
13-0 x 15-4

bath

up

BED RM./
STUDY
12-0 x 11-0

FOYER
5-0 x 13-6

DINING
12-0 x 13-2

PORCH
30-4 x 8-0

No. 99889
FIRST FLOOR PLAN

70-0

79-2

To order your Blueprints, call 1-800-235-5700

© 1995 Donald A. Gardner Architects, Inc.

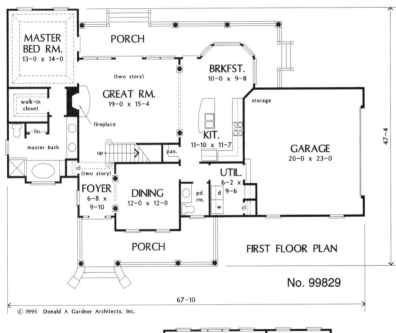

MASTER BED RM.
13-0 x 14-0

PORCH

BRKFST.
10-0 x 9-8

walk-in closet

GREAT RM.
19-0 x 15-4

(two story)

storage

lin.

fireplace

KIT.
11-10 x 11-7

GARAGE
20-0 x 23-0

master bath

up

pan.

47-4

UTIL.
6-2 x 9-6

cl
(two story)

FOYER
6-8 x 9-10

DINING
12-0 x 12-0

pd. rm.

d   cl
w

PORCH

FIRST FLOOR PLAN

No. 99829

© 1995 Donald A Gardner Architects, Inc.

67-10

TOTAL LIVING AREA:
1,972 SQ. FT.

great room below

BED RM.
11-10 x 13-0

skylights

cl   sto.

down

BONUS RM.
20-0 x 13-0

foyer below

sto.

lin.

BED RM.
12-0 x 12-0

walk-in closet

bath

attic storage

SECOND FLOOR PLAN

Distinctive Detailing

■ This plan features:

— Three bedrooms

— Two full and one half baths

■ Interior columns distinguish the inviting two-story Foyer from the Dining Room

■ Spacious Great Room set off by two-story windows and opening to the Kitchen and Breakfast Bay

■ Nine foot ceilings add volume and drama to the first floor

■ Secluded Master Suite topped by a space amplifying tray ceiling and enhanced by a plush bath

■ Two generous additional bedrooms with ample closet and storage space

■ Skylit bonus room enjoying second floor access

FIRST FLOOR — 1,436 SQ. FT.
SECOND FLOOR — 536 SQ. FT.
GARAGE & STORAGE — 520 SQ. FT.
BONUS ROOM — 296 SQ. FT.

Refer to **Pricing Schedule D** on the order form for pricing information

Functional Floor Plan

■ This plan features:

— Four bedrooms

— Three full baths

■ Columns accent front Porch and entrance into two-story Foyer

■ Living Room has hearth fireplace between French doors

■ Efficient Kitchen with peninsula counter, bright Breakfast area and adjoining Utility Room

■ Private Master Bedroom has a walk-in closet and plush bath with two vanities

■ First floor bedroom with walk-in closet and full bath access

■ Two additional bedrooms on second floor with dormers and large closets, share a full bath with separate vanities

■ No materials list available

■ Please specify a crawl space or slab foundation when ordering

FIRST FLOOR — 1,685 SQ. FT.
SECOND FLOOR — 648 SQ. FT.
GARAGE — 560 SQ. FT.

TOTAL LIVING AREA:
2,333 SQ. FT.

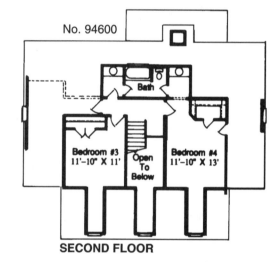

No. 94600

SECOND FLOOR

Bath

Bedroom #3
11'-10" X 11'

Open To Below

Bedroom #4
11'-10" X 13'

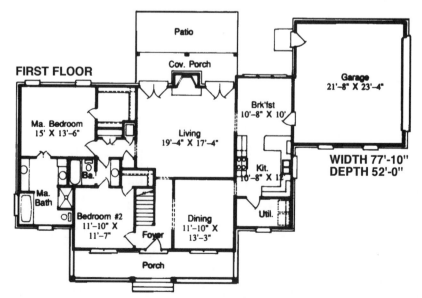

FIRST FLOOR

Patio

Cov. Porch

Garage
21'-8" X 23'-4"

Ma. Bedroom
15' X 13'-6"

Living
19'-4" X 17'-4"

Brk'fst
10'-8" X 10'

Ma. Bath

Bedroom #2
11'-10" X 11'-7"

Foyer

Kit.
10'-8" X 1.

Dining
11'-10" X 13'-3"

Util.

Porch

WIDTH 77'-10"
DEPTH 52'-0"

Refer to **Pricing Schedule B** on the order form for pricing information

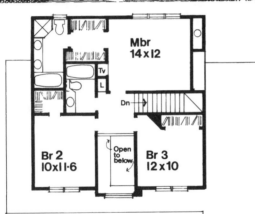

Mbr
14 x 12

Tv
L

Dn

Br 2
10 x 11·6

Open to below

Br 3
12 x 10

SECOND FLOOR

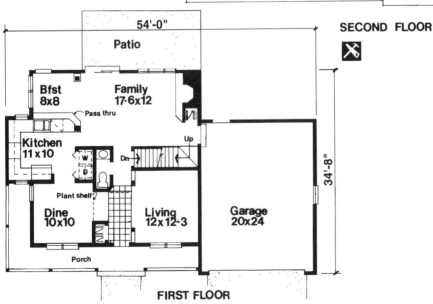

54'-0"

Patio

Bfst
8x8

Family
17·6x12

Pass thru

Kitchen
11 x 10

Up

W
D

Dn

Plant shelf

Dine
10x10

Living
12 x 12-3

Garage
20 x 24

34'-8"

Porch

FIRST FLOOR
No. 91903

Country Style Charmer

■ This plan features:

— Three bedrooms

— Two full and one half baths

■ A classical symmetry and gracious front porch

■ Formal areas zoned towards the front of the house

■ A large Family Room with fireplace

■ A convenient staircase located off the Family Room

■ A Master Bedroom with double vanity, separate glass shower and tub, and a built-in entertainment center

FIRST FLOOR — 910 SQ. FT.
SECOND FLOOR — 769 SQ. FT.
BASEMENT — 890 SQ. FT.

TOTAL LIVING AREA:
1,679 SQ. FT.

No. 20168
Farmhouse Flavor

■ This plan features:

— Three or four bedrooms

— Two full and one half baths

■ A Country style, three-sided porch, providing a warm welcome

■ A central staircase dominating the central Foyer, surrounded by the formal Living and Dining rooms

■ A decorative ceiling treatment adding elegance to the formal Dining Room

■ A quiet Den with built-in bookcases and a walk-in closet may also be used as a fourth bedroom

■ A spacious island Kitchen with a pantry, and a cheerful Breakfast Area with a decorative ceiling

■ A Sun Room highlighted by skylights and a plant shelf

■ A convenient first floor Master Suite with a skylight in the private bath

■ Two additional bedrooms, sharing a full bath loaded with intriguing angles

FIRST FLOOR — 1,698 SQ. FT.

SECOND FLOOR — 601 SQ. FT.

BASEMENT — 1,681 SQ. FT.

GARAGE — 616 SQ. FT.

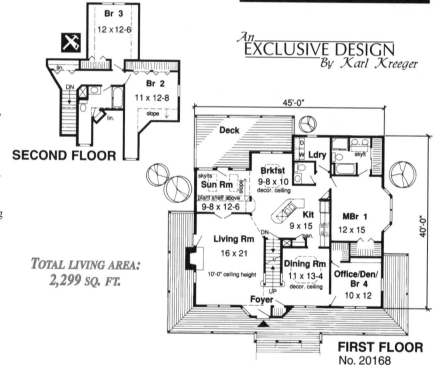

SECOND FLOOR

An
EXCLUSIVE DESIGN
By Karl Kreeger

TOTAL LIVING AREA:
2,299 SQ. FT.

FIRST FLOOR
No. 20168

No. 99888
Gracious Entrance

■ This plan features:

—Four bedrooms

—Three full and one half baths

■ Cathedral ceilings in the Foyer, Master Bath and Great Room.

■ Spacious Kitchen enhanced by an island with a built-in stove and a windowed Breakfast area overlooking a Porch with Spa.

■ Main level Master Suite with a walk-in closet, overlooking the back Porch

■ Utility/Laundry room between the oversized Garage and Kitchen.

■ Two bedrooms, a bathroom and a balcony overlooking the Great Room from the second floor

FIRST FLOOR — 2,238 SQ. FT.

SECOND FLOOR — 768 SQ. FT.

GARAGE — 865 SQ. FT.

TOTAL LIVING AREA:
3,006 SQ. FT.

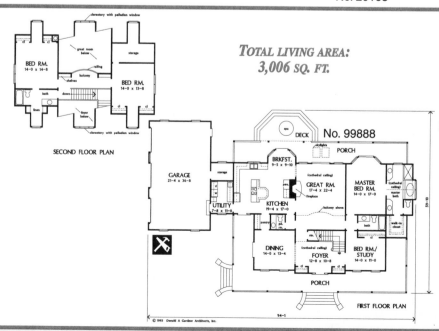

SECOND FLOOR PLAN

No. 99888

FIRST FLOOR PLAN

PRICE CODE F

No. 96444

Comfortable, Easy Living

■ This plan features:
— Five bedrooms
— Three full and one half baths

■ Great Room is overlooked by a curved balcony and features a fireplace with built-ins on either side

■ Spacious and efficient Kitchen equipped with a cooktop island has direct access to Breakfast Room and Dining Room

■ Swing Room, Bedroom/Study, with private full bath and closet

■ Master Suite pampered by lavish bath and large walk-in closet

■ Three additional second floor bedrooms, each with ample storage, share a full bath in the hall

FIRST FLOOR — 2,176 SQ. FT.
SECOND FLOOR — 861 SQ. FT.
BONUS ROOM — 483 SQ. FT.
GARAGE — 710 SQ. FT.

TOTAL LIVING AREA:
3,037 SQ. FT.

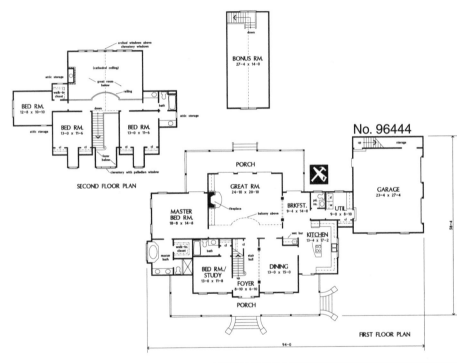

PRICE CODE F

PRICE CODE D

No. 98745
A Home to Grow in

■ This plan features:
— Three bedrooms
— Two full and one half baths
■ A covered veranda that sweeps around to the side of the house
■ A cozy Family Room open to the Dining Room and the Kitchen
■ A formal Dining area efficiently served by the Kitchen
■ A charming Master Suite enhanced by a private bath and a huge walk-in closet
■ Two additional bedrooms served by a full bath in the hall
■ Mud/Utility Room has outside entrance and easy access from garage

FIRST FLOOR — 1,830 SQ. FT.
SECOND FLOOR — 582 SQ. FT.
GARAGE — 572 SQ. FT.

TOTAL LIVING AREA:
2,412 SQ. FT.

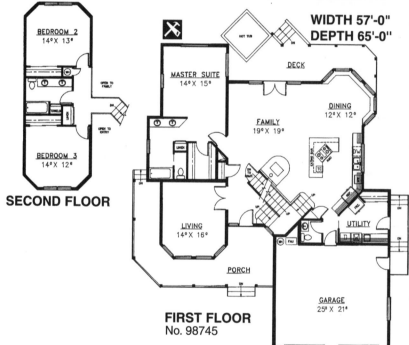

WIDTH 57'-0"
DEPTH 65'-0"

SECOND FLOOR
BEDROOM 2 14⁰X 13⁸
BEDROOM 3 14⁰X 12⁰

FIRST FLOOR
No. 98745
MASTER SUITE 14⁰X 15⁰
DECK
DINING 12⁰X 12⁰
FAMILY 19⁰X 19⁰
LIVING 14⁰X 16⁰
UTILITY
PORCH
GARAGE 25⁰X 21⁸
HOT TUB

No. 92269
Quality Inside and Out

■ This plan features:
— Three bedrooms
— Two full and one half baths
■ Decorative brick treatment adds to attractive facade and appeal
■ A spacious Entry/Gallery opens to Living Room with one of two fireplaces
■ Country Kitchen with work island opens to Breakfast bay, Family Room, Dining Room, Utility and Garage entry
■ Expansive Family Room offers a vaulted ceiling, third fireplace and access to Patio
■ Private Master Bedroom suite with a vaulted ceiling, lavish bath and walk-in closet
■ Two second floor bedrooms share a double vanity bath
■ No materials list available

MAIN FLOOR — 2,273 SQ. FT.
UPPER FLOOR — 562 SQ. FT.
GARAGE — 460 SQ. FT.

TOTAL LIVING AREA:
2,835 SQ. FT.

Upper Floor
Bed #3 13x14
Bed #2 12x14
B #2

Main Floor
No. 92269
73' - 0"
62' - 10"
Pool
Patio
FamilyRm 16x20
Brkfst
Bar
Kit 10x16
LivRm 17x17
MstrBed 16x17
Gallery
Util
Pwdr
Master
Ent
FmlDin 12x13
Study 12x14
Gar 20x23
Por

No. 99661

One-Story Farmhouse

■ This plan features:
— Three bedrooms
— Two full baths

■ Main activity space grouped to the right of the Foyer

■ Large Living Room with a corner fireplace and a front-facing bow window

■ Dining room enhanced by three French doors to the rear deck

■ Eat-in Kitchen directly accesses the Dining Room

■ Two bedrooms off a short hall sharing a full bath

■ Master Bedroom Suite with a vaulted ceiling, a window with an elliptical top, a walk-in closet and a private bath

MAIN FLOOR — 1,387 SQ. FT.
BASEMENT — 1,387 SQ. FT.
GARAGE — 493 SQ. FT.

TOTAL LIVING AREA:
1,387 SQ. FT.

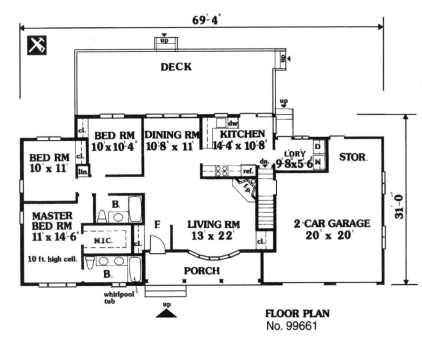

FLOOR PLAN
No. 99661

©1993 Donald A. Gardner Architects, Inc.

PRICE CODE D

No. 96446
Perfect for the Growing Family

■ This plan features:
— Three bedrooms
— Two full and one half baths
■ Natural light fills the two-story foyer through a palladian window in dormer above
■ Dining Room and Great Room adjoin for entertaining possibilities
■ U-shaped Kitchen with a curved counter opens to a large Breakfast Area
■ Master Suite, situated downstairs for privacy with generous walk-in closet, double vanity, separate shower and a whirlpool tub
■ Please specify a basement or crawl space foundation when ordering

FIRST FLOOR — 1,484 SQ. FT.
SECOND FLOOR — 660 SQ. FT.
BONUS ROOM — 389 SQ. FT.
GARAGE — 600 SQ. FT.

TOTAL LIVING AREA:
2,144 SQ. FT.

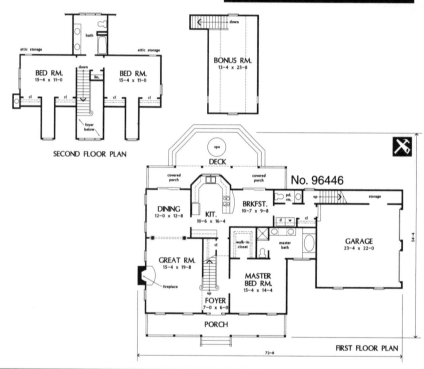

No. 98748
Mid-Sized Country Style

■ This plan features:
— Three bedrooms
— Two full baths
■ A vaulted ceiling over the Entry, Living Room, Dining Room, Family Room and Master Suite
■ Wide window bay expanding the Living Room
■ Wide garden window expanding the Kitchen while the cooktop, L-shaped island/eating bar adds convenience
■ A walk-in closet, oversized shower, a spa tub and a double vanity outside the bath area highlight the Master Suite
■ Two roomy additional bedrooms share the full bath in the hall

MAIN FLOOR — 2,126 SQ. FT.

TOTAL LIVING AREA:
2,126 SQ. FT.

WIDTH 64'-0"
DEPTH 64'-0"

No. 99862

Farmhouse with a Modern Flair

■ This plan features:

— Three bedrooms

— Two full and one half baths

■ Covered porch, arched windows, a two level Foyer and a bank of clerestory windows giving a modern flair to this farmhouse style

■ Columns between the generous Great Room and the island Kitchen

■ Large rear deck expanding the living space outdoors

■ Second level Master Suite with a walk-in closet, double vanity, shower and a garden tub

■ Two additional upstairs bedrooms sharing a full, skylit bath

■ Bonus room with skylights for future expansion

FIRST FLOOR — 943 SQ. FT.

SECOND FLOOR — 840 SQ. FT.

GARAGE & STORAGE — 510 SQ. FT.

BONUS ROOM — 323 SQ. FT.

TOTAL LIVING AREA:
1,783 SQ. FT.

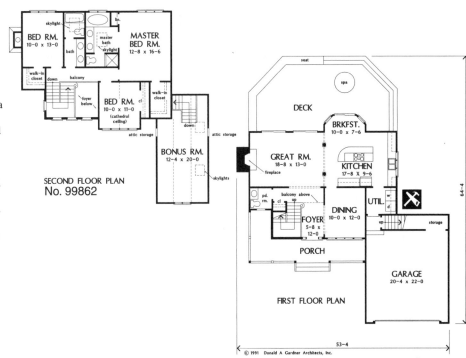

SECOND FLOOR PLAN
No. 99862

FIRST FLOOR PLAN

© 1991 Donald A Gardner Architects, Inc.

PRICE CODE D

No. 99796
Genteel Country Home

■ This plan features:
— Three bedrooms
— Two full and one half bath
■ A wrap-around porch and numerous Windows enhance the facade
■ A double door entry
■ A classic formal Living Room with a fireplace, built-in shelves and double doors for privacy
■ Vaulted ceilings in the Dining Room, Living Room and one Bedroom
■ A large first floor Master Suite with a lavish bath and a walk-in closet
■ A corner wood stove adding a cozy touch to the Family Room
■ A built-in wetbar dividing the Family Room from the Nook area
■ An efficient U-shaped Kitchen with a built-in planning desk, a peninsula counter/eating bar and easy garage entry
■ Two additional bedrooms on the second floor sharing a compartmented bath with two basins
■ Utility and Powder rooms located between the garage entry and the Kitchen

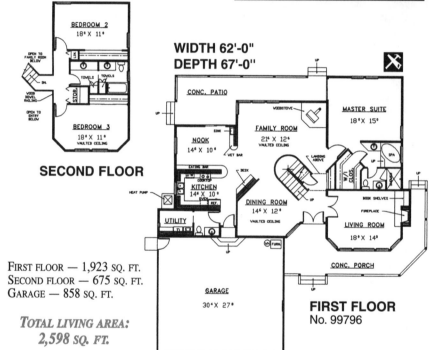

SECOND FLOOR

WIDTH 62'-0"
DEPTH 67'-0"

FIRST FLOOR
No. 99796

FIRST FLOOR — 1,923 SQ. FT.
SECOND FLOOR — 675 SQ. FT.
GARAGE — 858 SQ. FT.

TOTAL LIVING AREA:
2,598 SQ. FT.

No. 99055
Western Ranch with Spanish Influence

■ This plan features:
— Four bedrooms
— Two full and one half baths
■ Western styled covered front porch formed by tiled roof
■ Oval sunken living room with a mammoth stone fireplace and dramatic skylights
■ The curved end of oval Living Room surrounded by a wide terrace onto which living and dining rooms open
■ Efficient U-shaped Kitchen with extended counter
■ Master Bedroom highlighted by dressing rooms and private bath
■ Three additional bedrooms sharing full bath in hall

MAIN FLOOR — 1,800 SQ. FT.
BASEMENT — 1,780 SQ. FT.

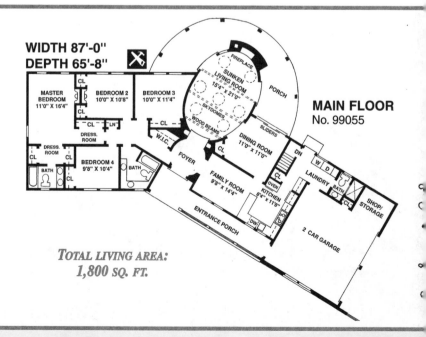

WIDTH 87'-0"
DEPTH 65'-8"

MAIN FLOOR
No. 99055

TOTAL LIVING AREA:
1,800 SQ. FT.

No. 93334
Attractive Curb Presence

■ This plan features:
— Four bedrooms
— Three full and one half baths

■ A Porch entrance leading into an open Foyer area with a Balcony and window seat above

■ An unusual pentagon-shaped Living Room topped by a stepped ceiling, and with French doors to a Sun Room for elegant entertaining

■ A stepped ceiling and a large bay window adding charm to the formal Dining Room

■ An informal Family Room with a massive fireplace, convenient built-ins, and a wall of windows, opening to Dinette/Kitchen area

■ An island Kitchen with an abundance of counter space, a double sink, built-in desk, pantry and a sky-lit Dinette

■ A luxurious Master Suite with a tray ceiling and a private Bath featuring a whirlpool, corner window tub, double vanity and an extra-large walk-in closet

■ On the second floor, three additional bedrooms, each with private access to a full bath, and a Bonus Room with many options

■ No materials list available

SECOND FLOOR

FIRST FLOOR — 1,970 SQ. FT.
SECOND FLOOR — 1,638 SQ. FT.
BONUS ROOM — 587 SQ. FT.

TOTAL LIVING AREA:
3,608 SQ. FT.

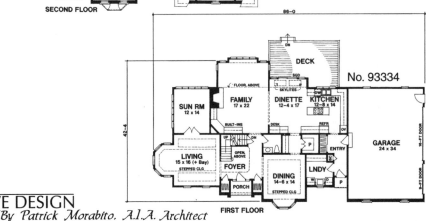

No. 93334

FIRST FLOOR

An EXCLUSIVE DESIGN
By Patrick Morabito, A.I.A. Architect

© 1992 Donald A. Gardner Architects, Inc.

No. 99890
Casually Elegant

- This plan features:
—Four bedrooms
—Three full and one half baths
- Two-level Foyer, naturally illuminated by the palladian window above
- Great Room topped by a cathedral ceiling and highlighted by a balcony above and a fireplace
- Columns defining the Great Room from the Kitchen and Breakfast Room
- First floor Master Suite opening to the screened porch through the bay area
- Two additional bedrooms on the second floor, each with a private bath

FIRST FLOOR — 1,766 SQ. FT.
SECOND FLOOR — 670 SQ. FT.
GARAGE & STORAGE — 624 SQ. FT.

TOTAL LIVING AREA:
2,436 SQ. FT.

SECOND FLOOR

BED RM. 12-8 x 11-6
LOFT 11-10 x 7-8
BED RM. 12-8 x 11-6

FIRST FLOOR
No. 99890

SCREEN PORCH 40-0 x 10-6
DECK
MASTER BED RM. 12-8 x 17-2
GREAT RM. 15-4 x 24-0
BRKFST. 10-4 x 8-8
UTILITY 9-6 x 9-8
GARAGE 23-4 x 21-8
KITCHEN 12-8 x 14-6
BED RM./ STUDY 12-8 x 11-0
FOYER 15-4 x 7-2
DINING 14-8 x 12-8
PORCH

© 1992 Donald A Gardner Architects, Inc.

No. 92217
Victorian Country Charm

- This plan features:
— Three bedrooms
— Two full and one half baths
- Curved front Porch leads into formal Entry
- Elegant, sloped ceiling crowns the formal Living Room
- Formal Dining Room adorned by an alcove of windows
- Open Kitchen with a cooktop/work island, bright Dinette area and nearby Utility room with Garage entry
- Family Room offers a cozy fireplace and access to Patio
- Expansive Master Bedroom highlighted by a cathedral ceiling, large walk-in closet and plush, skylit bath
- Two additional bedrooms have private access to a full bath
- No materials list available

MAIN FLOOR — 1,157 SQ. FT.
UPPER LEVEL — 838 SQ. FT.
GARAGE — 483 SQ. FT.

TOTAL LIVING AREA:
1,995 SQ. FT.

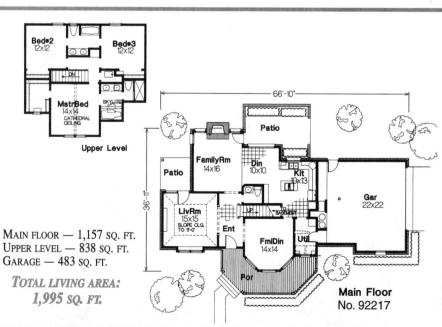

Bed#2 12x12
Bed#3 12x12
MstrBed 14x14 CATHEDRAL CEILING
SKYLITE

Upper Level

Patio
FamilyRm 14x16
Din 10x10
Kit 10x13
Patio
Gar 22x22
LivRm 15x15 SLOPE CLG. TO 17-0
BASEMENT
Ent
FmlDin 14x14
Util
Por

66'-10"
36'-11"

Main Floor
No. 92217

PRICE CODE C

No. 99873
Deck Includes Spa

- This plan features:
— Three bedrooms
— Two full and one half baths

- An exterior Porch giving the home a traditional flavor

- Great Room highlighted by a fireplace and a balcony above as well as a pass through into the kitchen

- Kitchen eating area with sky lights and bow windows overlooking the Deck with a spa

- Two additional bedrooms with a full bath on the second floor

- Master Suite on the first floor and naturally illuminated by two skylights

- Please specify a basement or crawl space foundation when ordering this plan

FIRST FLOOR — 1,325 SQ. FT.
SECOND FLOOR — 453 SQ. FT.

TOTAL LIVING AREA:
1,778 SQ. FT.

clerestory with palladian window

great room below

(cathedral ceiling)

railing

attic storage   attic storage

BED RM.   BED RM.
11-4 x 10-2   11-4 x 10-2

down   bath

cl   cl   cl   cl

attic storage   foyer below   attic storage

SECOND FLOOR PLAN
No. 99873

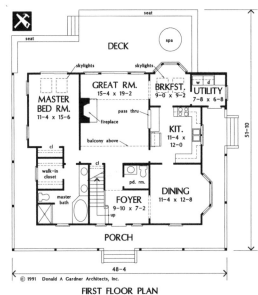

seat

DECK   spa

seat

skylights   skylights

GREAT RM.   BRKFST.   UTILITY
15-4 x 19-2   9-0 x 9-2   7-8 x 6-8

MASTER
BED RM.   pass thru
11-4 x 15-6   KIT.
                  fireplace   11-4 x
                               12-0
cl   balcony above

walk-in
closet   pd. rm.

                               DINING
                  cl   11-4 x 12-8
master
bath   FOYER
               9-10 x 7-2
               up

PORCH

48-4   51-10

© 1991 Donald A Gardner Architects, Inc.

FIRST FLOOR PLAN

PRICE CODE C

© 1991 Donald A. Gardner Architects, Inc.

No. 99893
Tradition With Flair

▪ This plan features:
— Four bedrooms
— Three full baths

▪ Spacious island Kitchen separating the Dining Room, Breakfast Area with skylights and the Utility Room

▪ Great Room enhanced by a fireplace and balcony above with direct access to the Deck with built-in seating and a spa

▪ Main floor Master Suite with two skylights, a walk-in closet and a spacious bath with unique windows

▪ Upstairs, two bedrooms sharing a dual vanity bathroom

FIRST FLOOR — 1,756 SQ. FT
SECOND FLOOR — 565 SQ. FT.

TOTAL LIVING AREA: 2,321 SQ. FT.

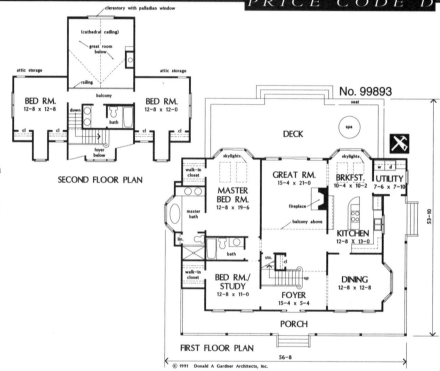

No. 96460
Open Spaces Compliment Flowing Design

▪ This plan features:
— Three bedrooms
— Two full and one half baths

▪ Dormers and a cathedral ceiling amplify the volume in the integrated Foyer and Great Room

▪ Kitchen, Dining Room, Breakfast Bay, and the Great Room access the back Porches

▪ Downstairs Master Bedroom accented by a tray ceiling with his-n-her walk-in closets

▪ Two additional bedrooms highlighted by a versatile nook for sitting or studying

▪ A full bath and a flexible skylit Bonus Room round out the plan

FIRST FLOOR — 1,458 SQ. FT.
SECOND FLOOR — 484 SQ. FT.
GARAGE & STORAGE — 497 SQ. FT.
BONUS ROOM — 257 SQ. FT.

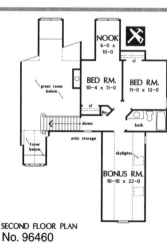

SECOND FLOOR PLAN
No. 96460

TOTAL LIVING AREA: 1,942 SQ. FT.

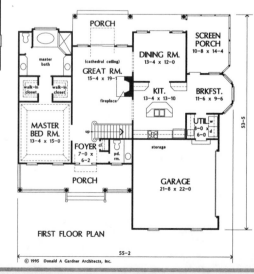

FIRST FLOOR PLAN

© 1995 Donald A Gardner Architects, Inc.

No. 99057

Spacious and
Convenient Ranch

■ This plan features:
— Three bedrooms
— One full and one three-quarter bath

■ Portico offers a sheltered entrance into Foyer and formal Living Room

■ Convenient Family Room with inviting fireplace, Patio access and nearby Laundry/Garage entry

■ Formal Dining Room highlighted by back yard view

■ U-shaped, efficient Kitchen with pantry and peninsula serving counter

■ Corner Master Bedroom offers three closets and a private bath

■ Two additional bedrooms share a full bath

MAIN FLOOR — 1,720 SQ. FT.

TOTAL LIVING AREA:
1,720 SQ. FT.

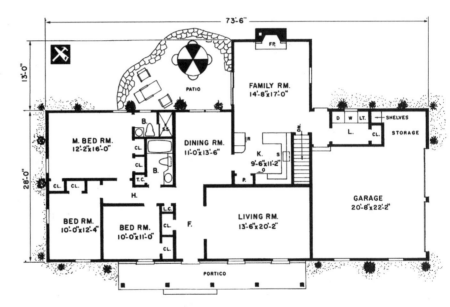

FIRST FLOOR PLAN
No. 99057

No. 92685
Compact One Level Home

■ This plan features:
— Three bedrooms
— Two full baths

■ Great Room combines with the Breakfast area to form a spacious gathering place

■ Sloped ceiling tops Great Room and reaches a twelve foot height

■ Windows across the rear of home provide a favorable indoor/outdoor relationship

■ Step-saving Kitchen with surplus counter space, cabinets and a pantry

■ Master Suite includes a walk-in closet and a full bath

■ Two additional bedrooms share the full bath in the hall

■ No materials list available

MAIN FLOOR — 1,442 SQ. FT.
BASEMENT — 1,442 SQ. FT.
GARAGE — 421 SQ. FT.

TOTAL LIVING AREA:
1,442 SQ. FT.

FIRST FLOOR
No. 92685

No. 99898
Curb Appeal

■ This plan features:
— Four bedrooms
— Three full and one half baths

■ Inviting front porch, dormers, gables, and windows topped by half rounds give this home curb appeal

■ An open floor plan with a split bedroom design and a spacious bonus room

■ Two dormers add light and volume to the Foyer

■ A cathedral ceiling enlarge the open Great Room

■ Accent columns define the Great Room, Kitchen, and Breakfast Area

■ Private Master Suite, with a tray ceiling and a lavish bath, accesses the rear deck through sliding glass doors

FIRST FLOOR — 2,920 SQ. FT.
SECOND FLOOR — 853 SQ. FT.
BONUS ROOM — 458 SQ. FT.
GARAGE — 680 SQ. FT.

TOTAL LIVING AREA:
3,773 SQ. FT.

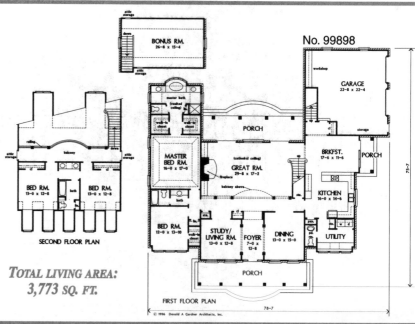

No. 99898

FIRST FLOOR PLAN

© 1996 Donald A Gardner Architects, Inc.

© 1994 Donald A. Gardner Architects, Inc.

B. NATHAN

No. 99846

Rambling Ranch

■ This plan features:
— Three bedrooms
— Two full and one half baths
■ Great Room is the focal point of home with its fireplace, built-in cabinets, cathedral ceiling, and access to screened porch and Kitchen
■ Privately situated Master Suite with skylights, a double vanity, a walk-in closet and access to the outside spa deck
■ Front of the house includes a large Foyer, a powder room and a formal Dining area
■ Two additional bedrooms share a full bath
■ The bonus room is complimented with skylights

MAIN FLOOR — 2,136 SQ. FT.
BONUS ROOM — 405 SQ. FT.
GARAGE — 670 SQ. FT.

TOTAL LIVING AREA:
2,136 SQ. FT.

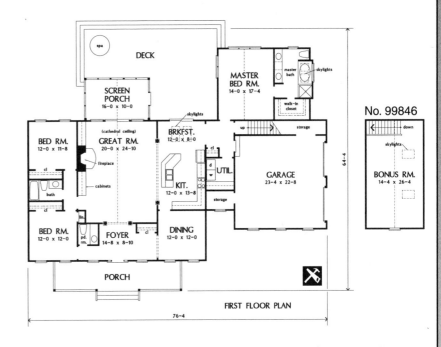

FIRST FLOOR PLAN

B. NATHAN

© 1996 Donald A. Gardner Architects, Inc.

PRICE CODE E

No. 99205
Farmhouse Feeling, Family-Style

■ This plan features:
— Four bedrooms
— Two full and two half baths

■ A sunny Breakfast bay with easy access to the efficient Kitchen

■ A large and spacious Family Room with a fireplace and a pass-through to the Kitchen

■ Sliders that link the Family and Dining Rooms with the rear terrace

■ A private Master Suite with his-and-her walk-in closets, dressing room with built-in vanity and convenient step-in shower

FIRST FLOOR — 1,590 SQ. FT.
SECOND FLOOR — 1,344 SQ. FT.
BASEMENT — 1,271 SQ. FT.

TOTAL LIVING AREA: 2,934 SQ. FT.

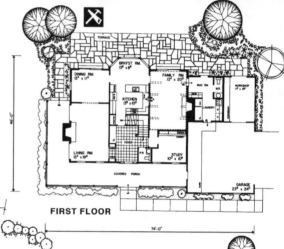

FIRST FLOOR

No. 99205

SECOND FLOOR

No. 99875
Wrapping Front Porch

■ This plan features:
— Three bedrooms
— Two full and one half baths

■ Two-story Foyer enjoys natural light from window above

■ Elegant bay window highlights the Dining Room

■ Kitchen directly accesses the formal Dining Room and the informal Breakfast Bay for ease in serving

■ Great Room accesses the skylit screened porch

■ Lavish Master Suite with garden tub, separate shower, double vanity and a walk-in closet

■ Two additional bedrooms on the second floor sharing a full double vanity bath in the hall

FIRST FLOOR — 1,526 SQ. FT.
SECOND FLOOR — 635 SQ. FT.
BONUS ROOM — 355 SQ. FT.

TOTAL LIVING AREA: 2,161 SQ. FT.

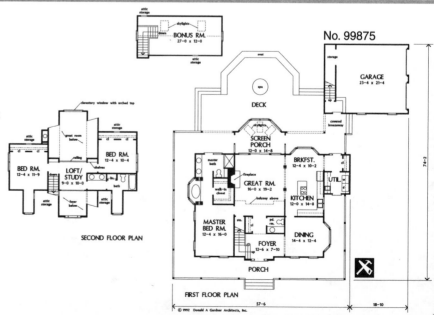

SECOND FLOOR PLAN

FIRST FLOOR PLAN

© 1992 Donald A Gardner Architects, Inc.

PRICE CODE F

No. 93338

Comfortable and Elegant Traditional

■ This plan features:
— Four bedrooms
— Three full and one half baths

■ Keystone arch frames entrance into Foyer with lovely landing staircase

■ Formal Living and Dining rooms provide ease in entertaining and views on three sides

■ Quiet Study with built-in book shelves offers many uses

■ Tray ceiling tops fireplace in large Family Room

■ Efficient, L-shaped Kitchen with cooktop work island, built-in desk and Dinette with access to Deck

■ Corner Master Bedroom offers a huge walk-in closet and plush bath retreat

■ Three additional bedrooms with access to full baths

■ No materials list available

FIRST FLOOR — 1,965 SQ. FT.
SECOND FLOOR — 1,781 SQ. FT.
BASEMENT — 1,965 SQ. FT.
GARAGE — 876 SQ. FT.

An EXCLUSIVE DESIGN
By Patrick Morabito, A.I.A. Architect

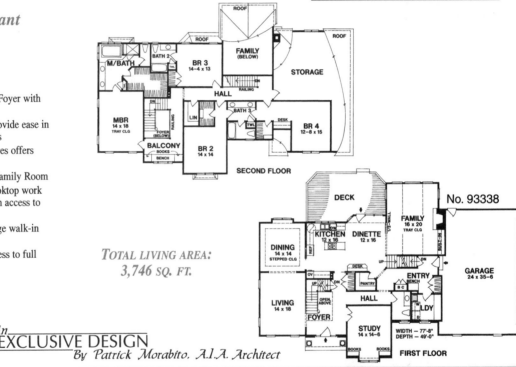

TOTAL LIVING AREA:
3,746 SQ. FT.

SECOND FLOOR

No. 93338

FIRST FLOOR

PRICE CODE D

PRICE CODE E

No. 99853
Stately Elegance

■ This plan features:

— Four bedrooms

— Three full and one half baths

■ Impressive double gable roof with front and rear palladian windows and wrap-around Porch

■ Vaulted ceilings in two-story Foyer and Great Room accommodates Loft/Study area

■ Spacious, first floor Master Bedroom Suite offers walk-in closet and luxurious bath

■ Living space expanded outdoors by wrap-around Porch and large Deck

■ Upstairs, one of three bedrooms could be a second master suite

FIRST FLOOR — 1,734 SQ. FT.

SECOND FLOOR — 958 SQ. FT.

TOTAL LIVING AREA:
2,692 SQ. FT.

SECOND FLOOR PLAN
No. 99853

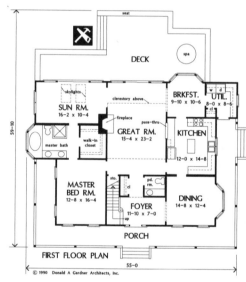

No. 99882
Relaxed Country Flair

■ This plan features:

— Three bedrooms

— Two full baths

■ Traditional stucco veneer combining with the relaxed country flair of a wrap-around porch

■ Two grand, double-door entrances offering easy access to the private Master Suite from the Great Room and rear porch

■ Bayed eating, a generous Great Room and unique, segmented hall bath

■ Sleeping quarters include the Master Bedroom with a vaulted ceiling, an additional family bedroom and a front bedroom/study

■ A popular U-shape Kitchen easily serving the Dining Room and the Breakfast area

MAIN FLOOR — 2,006 SQ. FT.

TOTAL LIVING AREA:
2,006 SQ. FT.

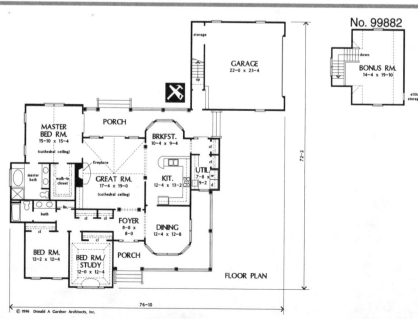

No. 99882

No. 99892
A Center Family Room

■ This plan features:
— Four bedrooms
— Three full and one half baths

■ At the center of the home is a fireplaced family room with a rounded balcony overlooking from above.

■ Master Suite enhanced by a sitting area with a unique window, a walk in closet and spacious bathroom with a dual vanity

■ Dining Room and Breakfast area both off the island Kitchen

■ A huge wrap-around porch extending to a deck, complete with built-in seating and a spa

■ A covered breezeway joining the garage to the utility/laundry room

FIRST FLOOR — 1,759 SQ. FT.
SECOND FLOOR — 888 SQ. FT.
GARAGE — 532 SQ. FT.

TOTAL LIVING AREA:
2,647 SQ. FT.

SECOND FLOOR PLAN

FIRST FLOOR PLAN

PRICE CODE C

No. 99260
Country Traditions

■ This plan features:
— Three bedrooms
— Two full baths

■ Cozy Covered Porch invites you into the Foyer

■ A windowed front-facing Breakfast Room

■ Efficient Kitchen with a corner Laundry Room, large Pantry, snack bar pass-through to the Gathering Room and an entry to the Dining Area

■ A massive Gathering Room and a Dining Room with an impressive fireplace and access to the rear Terrace

■ A whirlpool tub, separate shower and separate vanity area in the Master Suite

■ A Study at the front of the house that may be converted to a third bedroom

MAIN FLOOR — 1,835 SQ. FT.

TOTAL LIVING AREA:
1,835 SQ. FT.

WIDTH 71'-0"
DEPTH 43'-5"

MAIN FLOOR
No. 99260

No. 99850
Country on the Outside,
Contemporary on the Inside

■ This plan features:
— Three bedrooms
— Two full baths

■ Wrap-around porch and rear deck expanding living outdoors

■ Columns dramatically open and lead the foyer into a generous Great Room

■ Great Room open to the Kitchen/Breakfast area for a more spatial feeling

■ Natural light from the dormer windows flow into Foyer and Dining Room

■ Master Suite is privately located at the rear pampered by a private bath and a walk-in closet

■ Two front bedrooms share the full hall bath

MAIN FLOOR — 1,590 SQ. FT.
GARAGE & STORAGE — 506 SQ. FT.

TOTAL LIVING AREA:
1,590 SQ. FT.

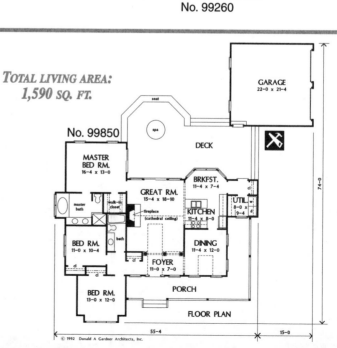

FLOOR PLAN

© 1992 Donald A Gardner Architects, Inc.

No. 92284

Fieldstone Facade

■ This plan features:
— Four bedrooms
— Two full and one half baths
■ Covered porch shelters entrance into Gallery and Great Room with a focal point fireplace and Patio access
■ Formal Dining Room conveniently located for entertaining
■ Cooktop island, built-in pantry and a bright Breakfast area highlight Kitchen
■ Secluded Master Bedroom suite with Patio access, large walk-in closet and corner spa tub
■ Three additional bedrooms with ample closets, share a double vanity bath
■ No materials list available

MAIN FLOOR — 2,261 SQ. FT.
GARAGE — 640 SQ. FT.

TOTAL LIVING AREA :
2,261 SQ. FT.

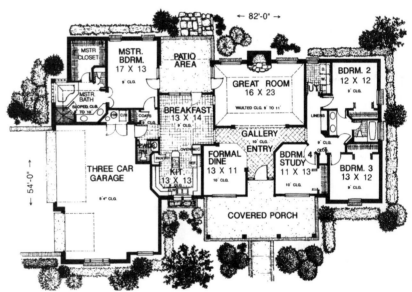

MAIN FLOOR
No. 92284

PRICE CODE C

No. 92218

*Victorian Details
Add Visual Delight*

■ This plan features:
— Three bedrooms
— Two full and one half baths

■ Quaint country porch provides a warm welcome

■ Expansive Living Room with a cozy fireplace and beamed ceiling

■ Formal Dining Room highlighted by an alcove of windows

■ An open Kitchen with cooktop work island, Dining area with Patio access and nearby Utility room with Garage entry

■ Comfortable Master Bedroom suite offers a cathedral ceiling, huge walk-in closet and a pampering bath

■ Two secondary bedrooms with ample closets and private access to a full bath

■ No materials list available

FIRST FLOOR — 1,082 SQ. FT.
SECOND FLOOR — 838 SQ. FT.
GARAGE — 500 SQ. FT.

TOTAL LIVING AREA:
1,920 SQ. FT.

SECOND FLOOR

FIRST FLOOR
No. 92218

No. 99860

Home Builders on a Budget

■ This plan features:
— Three bedrooms
— Two full baths

■ Down-sized country plan for home builder on a budget

■ Columns punctuate open, one-level floor plan and connect Foyer with clerestory window dormers

■ Front Porch and large, rear Deck extend living space outdoors

■ Tray ceilings decorate Master Bedroom, Dining Room and Bedroom/Study

■ Private Master Bath features garden tub, double vanity, separate shower and skylights

MAIN FLOOR — 1,498 SQ. FT.
GARAGE & STORAGE — 427 SQ. FT.

TOTAL LIVING AREA:
1,498 SQ. FT

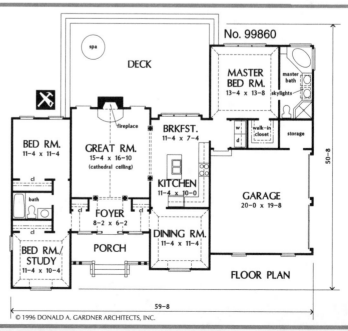

No. 99860

FLOOR PLAN

©1994 Donald A. Gardner Architects, Inc.

PRICE CODE D

No. 96451
*Wonderfully Livable,
Efficient Floor Plan*

■ This plan features:
— Three bedrooms
— Two full and one half baths

■ Foyer with accent columns opening to a large Dining Room and Great Room with a cathedral ceiling

■ Kitchen, half open, half private with an extended counter/snack bar

■ Master Suite tucked into rear right corner, topped by a cathedral ceiling, and highlighted by a fireplace, skylit tub, and separate shower

■ Bonus room over the garage for future expansion and storage

MAIN FLOOR — 2,123 SQ. FT.
GARAGE & STORAGE — 650 SQ. FT.
BONUS ROOM — 439 SQ. FT.

TOTAL LIVING AREA:
2,123 SQ. FT.

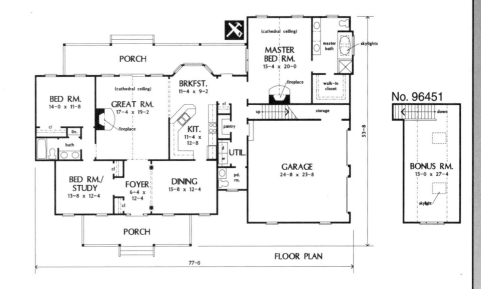

PORCH

BED RM.
14-0 x 11-8

GREAT RM.
17-4 x 19-2
(cathedral ceiling)
fireplace

BRKFST.
11-4 x 9-2

KIT.
11-4 x 12-8

pantry

UTIL.

(cathedral ceiling)

MASTER
BED RM.
15-4 x 20-0

master bath

skylights

fireplace

walk-in closet

up

storage

BED RM./
STUDY
13-8 x 12-4

FOYER
6-4 x 12-4

DINING
15-8 x 12-4

pd. rm.

GARAGE
24-8 x 23-8

PORCH

77-0

53-8

FLOOR PLAN

No. 96451

down

BONUS RM.
15-0 x 27-4

skylight

PRICE CODE B

© 1996 Donald A. Gardner Architects, Inc.

No. 96461
Country Classic

- ■ This plan features:
- — Three bedrooms
- — Two full and one half baths
- ■ Casually elegant exterior with dormers, gables and a charming front porch
- ■ U-shaped Kitchen easily serves both adjacent eating areas
- ■ Nine foot ceilings amplify the first floor
- ■ Master Suite highlighted by a vaulted ceiling and dormer
- ■ Garden tub with a double window are focus of the master bath
- ■ Two bedrooms with walk-in closets sharing a hall bath, while back stairs lead to a spacious bonus room

FIRST FLOOR — 1,313 SQ. FT.
SECOND FLOOR — 525 SQ. FT.
BONUS ROOM — 367 SQ. FT.
GARAGE — 513 SQ. FT.

TOTAL LIVING AREA:
1,838 SQ. FT.

SECOND FLOOR PLAN
No. 96461

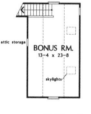

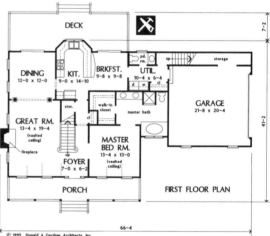

FIRST FLOOR PLAN

© 1995 Donald A Gardner Architects, Inc.

No. 93336
Executive Treatment

- ■ This plan features:
- — Four bedrooms
- — Three full and one half baths
- ■ Two high ceiling Entries with landing staircases connecting a Hall/Balcony
- ■ An elegant, formal Dining Room with pocket door to Kitchen area and steps to formal Living Room with inviting fireplace
- ■ A well-equipped Kitchen with a cooktop island/eating bar, a double sink, built-in pantry and desk, and a Dinette with sliding glass doors to the Screened Porch
- ■ A private Master Suite with an exclusive Screened Porch, dressing area, walk-in closet, a bath with a whirlpool tub, and Sitting Room
- ■ Three additional bedrooms on the second floor, each with private access to a full bath, and a Bonus Room with many options
- ■ No materials list available

No. 93336

FIRST FLOOR

SECOND FLOOR

FIRST FLOOR — 2,092 SQ. FT.
SECOND FLOOR — 1,934 SQ. FT.
BONUS — 508 SQ. FT.

TOTAL LIVING AREA:
4,026 SQ. FT.

An EXCLUSIVE DESIGN
By Patrick Morabito, A.I.A. Architect

No. 99240

A Master Suite To Love

■ This plan features:
— Three bedrooms
— Two full and two half baths

■ A large Sun Room with a cathedral ceiling and floor-to-ceiling windows

■ An enormous country Kitchen with a center range, double wall ovens, more than ample counter space, and a snack bar

■ A Clutter Room with a walk-in pantry, half-bath, and laundry/work area

■ A double-sided fireplace that opens the Kitchen to the Living Room with a wall-length raised hearth

■ A huge Master Suite with his-n-her walk-in closets, dressing or exercise room and private Master Bath

FIRST FLOOR — 3,511 SQ. FT.
SECOND FLOOR — 711 SQ. FT.
GARAGE — 841 SQ. FT.
WORKSHOP/THIRD GARAGE BAY —
231 SQ. FT.

TOTAL LIVING AREA:
4,222 SQ. FT.

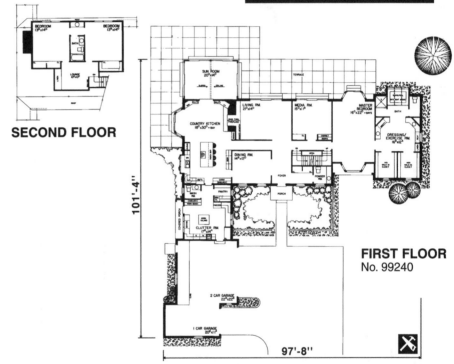

SECOND FLOOR

FIRST FLOOR
No. 99240

Everything You Need...
...to Make Your Dream Come True!

You pay only a fraction of the original cost for home designs by respected professionals.

You've Picked Your Dream Home!

You can already see it standing on your lot... you can see yourselves in your new home... enjoying family, entertaining guests, celebrating holidays. All that remains ahead are the details. That's where we can help. Whether you plan to build-it-yourself, be your own contractor, or hand your plans over to an outside contractor, your Garlinghouse blueprints provide the perfect beginning for putting yourself in your dream home right away.

We even make it simple for you to make professional design modifications. We can also provide a materials list for greater economy.

My grandfather, L.F. Garlinghouse, started a tradition of quality when he founded this company in 1907. For over 90 years, homeowners and builders have relied on us for accurate, complete, professional blueprints. Our plans help you get results fast... and save money, too! These pages will give you all the information you need to order. So get started now... I know you'll love your new Garlinghouse home!

Sincerely,

White Garling

EXTERIOR ELEVATIONS

Elevations are scaled drawings of the front, rear, left and right sides of a home. All of the necessary information pertaining to the exterior finish materials, roof pitches and exterior height dimensions of your home are defined.

CABINET PLANS

These plans, or in some cases elevations, will detail the layout of the kitchen and bathroom cabinets at a larger scale. This gives you an accurate layout for your cabinets or an ideal starting point for a modified custom cabinet design.

TYPICAL WALL SECTION

This section is provided to help your builder understand the structural components and materials used to construct the exterior walls of your home. This section will address insulation, roof components, and interior and exterior wall finishes. Your plans will be designed with either 2x4 or 2x6 exterior walls, but most professional contractors can easily adapt the plans to the wall thickness you require.

FIREPLACE DETAILS

If the home you have chosen includes a fireplace, the fireplace detail will show typical methods to construct the firebox, hearth and flue chase for masonry units, or a wood frame chase for a zero-clearance unit.

FOUNDATION PLAN

These plans will accurately dimension the footprint of your home including load bearing points and beam placement if applicable. The foundation style will vary from plan to plan. Your local climatic conditions will dictate whether a basement, slab or crawlspace is best suited for your area. In most cases, if your plan comes with one foundation style, a professional contractor can easily adapt the foundation plan to an alternate style.

ROOF PLAN

The information necessary to construct the roof will be included with your home plans. Some plans will reference roof trusses, while many others contain schematic framing plans. These framing plans will indicate the lumber sizes necessary for the rafters and ridgeboards based on the designated roof loads.

TYPICAL CROSS SECTION

A cut-away cross-section through the entire home shows your building contractor the exact correlation of construction components at all levels of the house. It will help to clarify the load bearing points from the roof all the way down to the basement.

DETAILED FLOOR PLANS

The floor plans of your home accurately dimension the positioning of all walls, doors, windows, stairs and permanent fixtures. They will show you the relationship and dimensions of rooms, closets and traffic patterns. Included is the schematic of the electrical layout. This layout is clearly represented and does not hinder the clarity of other pertinent information shown. All these details will help your builder properly construct your new home.

STAIR DETAILS

If stairs are an element of the design you have chosen, then a cross-section of the stairs will be included in your home plans. This gives your builders the essential reference points that they need for headroom clearance, and riser and tread dimensions.

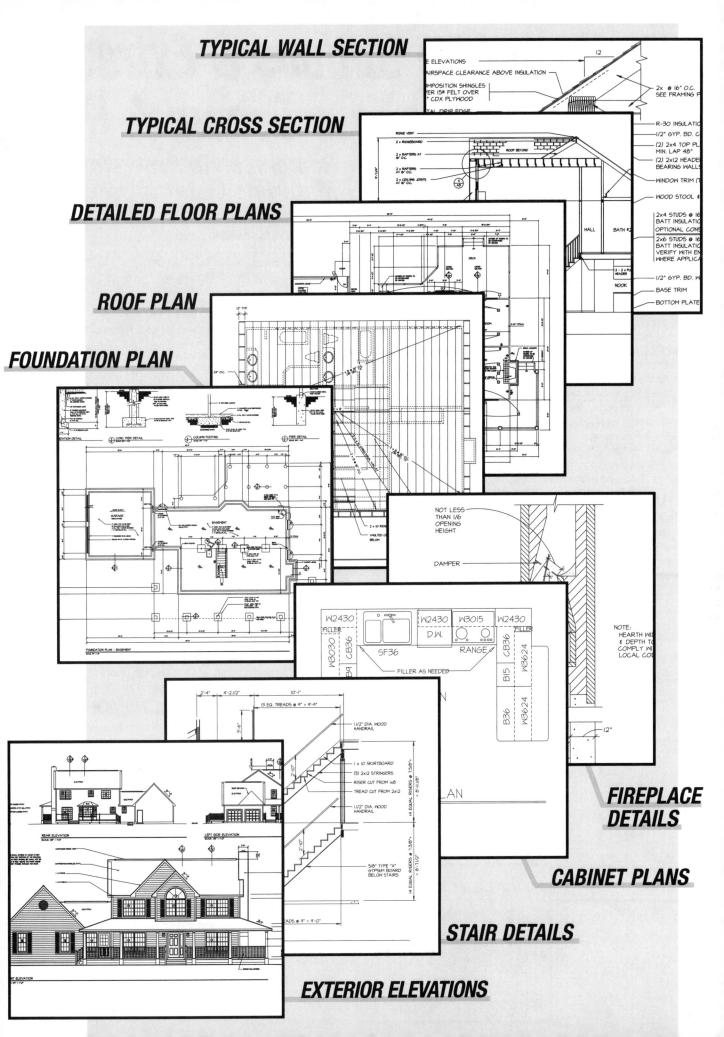

TYPICAL WALL SECTION

TYPICAL CROSS SECTION

DETAILED FLOOR PLANS

ROOF PLAN

FOUNDATION PLAN

FIREPLACE DETAILS

CABINET PLANS

STAIR DETAILS

EXTERIOR ELEVATIONS

Garlinghouse Options & Extras
...Make Your Dream A Home

Reversed Plans Can Make Your Dream Home Just Right!

"That's our dream home...if only the garage were on the other side!"

You could have exactly the home you want by flipping it end-for-end. Check it out by holding your dream home page of this book up to a mirror. Then simply order your plans "reversed." We'll send you one full set of mirror-image plans (with the writing backwards) as a master guide for you and your builder.

The remaining sets of your order will come as shown in this book so the dimensions and specifications are easily read on the job site...but most plans in our collection come stamped "REVERSED" so there is no construction confusion.

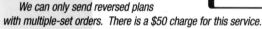

As Shown   Reversed

We can only send reversed plans with multiple-set orders. There is a $50 charge for this service.

Some plans in our collection are available in Right Reading Reverse. Right Reading Reverse plans will show your home in reverse, with the writing on the plan being readable. This easy-to-read format will save you valuable time and money. Please contact our Customer Service Department at (860) 343-5977 to check for Right Reading Reverse availability. (There is a $125 charge for this service.)

Specifications & Contract Form

We send this form to you free of charge with your home plan order. The form is designed to be filled in by you or your contractor with the exact materials to use in the construction of your new home. Once signed by you and your contractor it will provide you with peace of mind throughout the construction process.

$19.95 per set
(includes postage)

Remember To Order Your Materials List

It'll help you save money. Available at a modest additional charge, the Materials List gives the quantity, dimensions, and specifications for the major materials needed to build your home. You will get faster, more accurate bids from your contractors and building suppliers — and avoid paying for unused materials and waste. Materials Lists are available for all home plans except as otherwise indicated, but can only be ordered with a set of home plans. Due to differences in regional requirements and homeowner or builder preferences... electrical, plumbing and heating/air conditioning equipment specifications are not designed specifically for each plan. However, non-plan specific detailed typical prints of residential electrical, plumbing and construction guidelines can be provided. Please see below for additional information. If you need a detailed materials cost you might need to purchase a Zip Quote. (Details follow)

Detail Plans Provide Valuable Information About Construction Techniques

Because local codes and requirements vary greatly, we recommend that you obtain drawings and bids from licensed contractors to do your mechanical plans. However, if you want to know more about techniques — and deal more confidently with subcontractors — we offer these remarkably useful detail sheets. These detail sheets will aid in your understanding of these technical subjects. **The detail sheets are not specific to any one home plan and should be used only as a general reference guide.**

RESIDENTIAL CONSTRUCTION DETAILS

Ten sheets that cover the essentials of stick-built residential home construction. Details foundation options — poured concrete basement, concrete block, or monolithic concrete slab. Shows all aspects of floor, wall and roof framing. Provides details for roof dormers, overhangs, chimneys and skylights. Conforms to requirements of Uniform Building code or BOCA code. Includes a quick index and a glossary of terms.

RESIDENTIAL PLUMBING DETAILS

Eight sheets packed with information detailing pipe installation methods, fittings, and sized. Details plumbing hook-ups for toilets, sinks, washers, sump pumps, and septic system construction. Conforms to requirements of National Plumbing code. Color coded with a glossary of terms and quick index.

RESIDENTIAL ELECTRICAL DETAILS

Eight sheets that cover all aspects of residential wiring, from simple switch wiring to service entrance connections. Details distribution panel layout with outlet and switch schematics, circuit breaker and wiring installation methods, and ground fault interrupter specifications. Conforms to requirements of National Electrical Code. Color coded with a glossary of terms.

Modifying Your Favorite Design, Made EASY!

Modifying Your Garlinghouse Home Plan

Simple modifications to your dream home, including minor non-structural changes and material substitutions, can be made between you and your builder by marking the changes directly on your blueprints. However, if you are considering making significant changes to your chosen design, we recommend that you use the services of The Garlinghouse Co. Design Staff. We will help take your ideas and turn them into a reality, just the way you want. Here's our procedure!

When you place your Vellum order, you may also request a free Garlinghouse Modification Kit. In this kit, you will receive a red marking pencil, furniture cut-out sheet, ruler, a self addressed mailing label and a form for specifying any additional notes or drawings that will help us understand your design ideas. Mark your desired changes directly on the Vellum drawings. NOTE: Please use only a **red pencil** to mark your desired changes on the Vellum. Then, return the redlined Vellum set in the original box to The Garlinghouse Company at, 282 Main Street Extension, Middletown, CT 06457. **IMPORTANT:** Please **roll** the Vellums for shipping, **do not fold** the Vellums for shipping.

We also offer modification estimates. We will provide you with an estimate to draft your changes based on your specific modifications before you purchase the vellums, for a $50 fee. After you receive your estimate, if you decide to have The Garlinghouse Company Design Staff do the changes, the $50 estimate fee will be deducted from the cost of your modifications. If, however, you choose to use a different service, the $50 estimate fee is non-refundable.

Within 5 days of receipt of your plans, you will be contacted by a member of The Garlinghouse Co. Design Staff with an estimate for the design services to draw those changes. A 50% deposit is required before we begin making the actual modifications to your plans.

Once the preliminary design changes have been made to the floor plans and elevations, copies will be sent to you to make sure we have made the exact changes you want. We will wait for your approval before continuing with any structural revisions. The Garlinghouse Co. Design Staff will call again to inform you that your modified Vellum plan is complete and will be shipped as soon as the final payment has been made. For additional information call us at 1-860-343-5977. Please refer to the Modification Pricing Guide for estimated modification costs. Please call for Vellum modification availability for plan numbers 85,000 and above.

Reproducible Vellums for Local Modification Ease

If you decide not to use the Garlinghouse Co. Design Staff for your modifications, we recommend that you follow our same procedure of purchasing our Vellums. You then have the option of using the services of the original designer of the plan, a local professional designer, or architect to make the modifications to your plan.

With a Vellum copy of our plans, a design professional can alter the drawings just the way you want, then you can print as many copies of the modified plans as you need to build your house. And, since you have already started with our complete detailed plans, the cost of those expensive professional services will be significantly less than starting from scratch. Refer to the price schedule for Vellum costs. Again, please call for Vellum availability for plan numbers 85,000 and above.

IMPORTANT RETURN POLICY: Upon receipt of your Vellums, if for some reason you decide you do not want a modified plan, then simply return the Kit and the unopened Vellums. Reproducible Vellum copies of our home plans are copyright protected and only sold under the terms of a license agreement that you will receive with your order. Should you not agree to the terms, then the Vellums may be returned, **unopened,** for a full refund less the shipping and handling charges, plus a 15% restocking fee. For any additional information, please call us at 1-860-343-5977.

MODIFICATION PRICING GUIDE

CATEGORIES	ESTIMATED COST
KITCHEN LAYOUT — PLAN AND ELEVATION	$175.00
BATHROOM LAYOUT — PLAN AND ELEVATION	$175.00
FIREPLACE PLAN AND DETAILS	$200.00
INTERIOR ELEVATION	$125.00
EXTERIOR ELEVATION — MATERIAL CHANGE	$140.00
EXTERIOR ELEVATION — ADD BRICK OR STONE	$400.00
EXTERIOR ELEVATION — STYLE CHANGE	$450.00
NON BEARING WALLS (INTERIOR)	$200.00
BEARING AND/OR EXTERIOR WALLS	$325.00
WALL FRAMING CHANGE — 2X4 TO 2X6 OR 2X6 TO 2X4	$240.00
ADD/REDUCE LIVING SPACE — SQUARE FOOTAGE	QUOTE REQUIRED
NEW MATERIALS LIST	$.20 SQUARE FOOT
CHANGE TRUSSES TO RAFTERS OR CHANGE ROOF PITCH	$300.00
FRAMING PLAN CHANGES	$325.00
GARAGE CHANGES	$325.00
ADD A FOUNDATION OPTION	$300.00
FOUNDATION CHANGES	$250.00
RIGHT READING PLAN REVERSE	$575.00
ARCHITECTS SEAL	$300.00
ENERGY CERTIFICATE	$150.00
LIGHT AND VENTILATION SCHEDULE	$150.00

Questions?

Call our customer service department at **1-860-343-5977**

ZIP-QUOTE!
HOME COST CALCULATOR

ZIP QUOTE
HOME COST CALCULATOR

WHY?

Do you wish you could quickly find out the building cost for your new home without waiting for a contractor to compile hundreds of bids? Would you like to have a benchmark to compare your contractor(s) bids against? **Well, Now You Can!!,** with **Zip-Quote** Home Cost Calculator. Zip-Quote is only available for zip code areas within the United States.

HOW?

Our new **Zip-Quote** Home Cost Calculator will enable you to obtain the calculated building cost to construct your new home, based on labor rates and building material costs within your zip code area, without the normal delays or hassles usually associated with the bidding process. Zip-Quote can be purchased in two separate formats, an itemized or a bottom line format.

Zip-Quote is available for plans where you see this symbol.

"How does **Zip-Quote** actually work?" When we receive your **Zip-Quote** order, we process your specific home plan building materials list through our Home Cost Calculator which contains up-to-date rates for all residential labor trades and building material costs in your zip code area. "The result?" A calculated cost to build your dream home in your zip code area. This calculation will help you (as a consumer or a builder) evaluate your building budget. This is a valuable tool for anyone considering building a new home.

All database information for our calculations is furnished by Marshall & Swift, L.P. For over 60 years, Marshall & Swift L.P. has been a leading provider of cost data to professionals in all aspects of the construction and remodeling industries.

OPTION 1

The **Itemized Zip-Quote** is a detailed building material list. Each building material list line item will separately state the labor cost, material cost and equipment cost (if applicable) for the use of that building material in the construction process. Each category within the building material list will be subtotaled and the entire Itemized cost calculation totaled at the end. This building materials list will be summarized by the individual building categories and will have additional columns where you can enter data from your contractor's estimates for a cost comparison between the different suppliers and contractors who will actually quote you their products and services.

OPTION 2

The **Bottom Line Zip-Quote** is a one line summarized total cost for the home plan of your choice. This cost calculation is also based on the labor cost, material cost and equipment cost (if applicable) within your local zip code area.

COST

The price of your **Itemized Zip-Quote** is based upon the pricing schedule of the plan you have selected, in addition to the price of the materials list. Please refer to the pricing schedule on our order form. The price of your initial **Bottom Line Zip-Quote** is $29.95. Each additional **Bottom Line Zip-Quote** ordered in conjunction with the initial order is only $14.95. **Bottom Line Zip-Quote** may be purchased separately and does NOT have to be purchased in conjunction with a home plan order.

FYI

An **Itemized Zip-Quote** Home Cost Calculation can ONLY be purchased in conjunction with a Home Plan order. The **Itemized Zip-Quote** can not be purchased separately. The **Bottom Line Zip-Quote** can be purchased seperately and doesn't have to be purchased in conjunction with a home plan order. Please consult with a sales representative for current availability. If you find within 60 days of your order date that you will be unable to build this home, then you may exchange the plans and the materials list towards the price of a new set of plans (see order info pages for plan exchange policy). The **Itemized Zip-Quote** and the **Bottom Line Zip-Quote** are NOT returnable. The price of the initial **Bottom Line Zip-Quote** order can be credited towards the purchase of an **Itemized Zip-Quote** order only. Additional **Bottom Line Zip-Quote** orders, within the same order can not be credited. Please call our Customer Service Department for more information.

SOME MORE INFORMATION

The Itemized and Bottom Line Zip-Quotes give you approximated costs for constructing the particular house in your area. These costs are not exact and are only intended to be used as a preliminary estimate to help determine the affordability of a new home and/or as a guide to evaluate the general competitiveness of actual price quotes obtained through local suppliers and contractors. However, Zip-Quote cost figures should never be relied upon as the only source of information in either case. The Garlinghouse Company and Marshall & Swift L.P. can not guarantee any level of data accuracy or correctness in a Zip-Quote and disclaim all liability for loss with respect to the same, in excess of the original purchase price of the Zip-Quote product. All Zip-Quote calculations are based upon the actual blueprint materials list with options as selected by customer and do not reflect any differences that may be shown on the published house renderings, floor plans, or photographs.

Ignoring Copyright Laws Can Be
A $1,000,000 Mistake

Recent changes in the US copyright laws allow for statutory penalties of up to **$100,000** per incident for copyright infringement involving any of the copyrighted plans found in this publication. The law can be confusing. So, for your own protection, take the time to understand what you can and cannot do when it comes to home plans.

···WHAT YOU CANNOT DO ···

You Cannot Duplicate Home Plans

Purchasing a set of blueprints and making additional sets by reproducing the original is **illegal**. If you need multiple sets of a particular home plan, then you must purchase them.

You Cannot Copy Any Part of a Home Plan to Create Another

Creating your own plan by copying even part of a home design found in this publication is called "creating a derivative work" and is **illegal** unless you have permission to do so.

You Cannot Build a Home Without a License

You must have specific permission or license to build a home from a copyrighted design, even if the finished home has been changed from the original plan. It is **illegal** to build one of the homes found in this publication without a license.

What Garlinghouse Offers

Home Plan Blueprint Package

By purchasing a single or multiple set package of blueprints from Garlinghouse, you not only receive the physical blueprint documents necessary for construction, but you are also granted a license to build one, and only one, home. You can also make simple modifications, including minor non-structural changes and material substitutions, to our design, as long as these changes are made directly on the blueprints purchased from Garlinghouse and no additional copies are made.

Home Plan Vellums

By purchasing vellums for one of our home plans, you receive the same construction drawings found in the blueprints, but printed on vellum paper. Vellums can be erased and are perfect for making design changes. They are also semi-transparent making them easy to duplicate. But most importantly, the purchase of home plan vellums comes with a broader license that allows you to make changes to the design (ie, create a hand drawn or CAD derivative work), to make an unlimited number of copies of the plan, and to build one home from the plan.

License To Build Additional Homes

With the purchase of a blueprint package or vellums you automatically receive a license to build one home and only one home, respectively. If you want to build more homes than you are licensed to build through your purchase of a plan, then additional licenses may be purchased at reasonable costs from Garlinghouse. Inquire for more information.

Order Form

Order Code No. H8CH3

Plan prices guaranteed until 1/30/99 —After this date call for updated pricing

____ set(s) of blueprints for plan #_____	$_____
____ Vellum & Modification kit for plan #_____	$_____
____ Additional set(s) @ $30 each for plan #_____	$_____
____ Mirror Image Reverse @ $50 each	$_____
____ Right Reading Reverse @ $125 each	$_____
____ Materials list for plan #_____	$_____
____ Detail Plans @ $19.95 each	
❏ Construction ❏ Plumbing ❏ Electrical	$_____
____ Bottom line ZIP Quote@$29.95 for plan #_____	$_____
____ Additional Bottom Line Zip Quote	
@ $14.95 for plan(s) #_____	
_____	$_____
____ Itemized ZIP Quote for plan(s) #_____	$_____
Shipping (see charts on opposite page)	$_____
Subtotal	$_____
Sales Tax(CT residents add 6% sales tax, KS residents add 6.15% sales tax) (Not required for all states)	$_____
TOTAL AMOUNT ENCLOSED	**$_____**

Send your check, money order or credit card information to:
(No C.O.D.'s Please)

Please submit all <u>United States</u> & <u>Other Nations</u> orders to:

Garlinghouse Company
P.O. Box 1717
Middletown, CT. 06457

Please Submit all <u>Canadian</u> plan orders to:

Garlinghouse Company
60 Baffin Place, Unit #5
Waterloo, Ontario N2V 1Z7

ADDRESS INFORMATION:

NAME:_____

STREET:_____

CITY:_____

STATE:_____ **ZIP:**_____

DAYTIME PHONE:_____

Credit Card Information

Charge To:	❏ Visa	❏ Mastercard

Card # | | | | | | | | | | | | | | | | |

Signature _____ Exp. _____ / _____

IMPORTANT INFORMATION TO READ BEFORE YOU PLACE YOUR ORDER

How Many Sets Of Plans Will You Need?

The Standard 8-Set Construction Package

Our experience shows that you'll speed every step of construction and avoid costly building errors by ordering enough sets to go around. Each tradesperson wants a set — the general contractor and all subcontractors; foundation, electrical, plumbing, heating/air conditioning and framers. Don't forget your lending institution, building department and, of course, a set for yourself.

The Minimum 4-Set Construction Package

If you're comfortable with arduous follow-up, this package can save you a few dollars by giving you the option of passing down plan sets as work progresses. You might have enough copies to go around if work goes exactly as scheduled and no plans are lost or damaged by subcontractors. But for only $50 more, the 8-set package eliminates these worries.

The Single Study Set

We offer this set so you can study the blueprints to plan your dream home in detail. As with all of our plans, they are stamped with a copyright warning. Remember, one set is never enough to build your home. In pursuant to copyright laws, it is <u>illegal</u> to reproduce any blueprint.

Our Reorder and Exchange Policies:

If you find after your initial purchase that you require additional sets of plans you may purchase them from us at special reorder prices (please call for pricing details) provided that you reorder within 6 months of your original order date. There is a $28 reorder processing fee that is charged on all reorders. For more information on reordering plans please contact our Customer Service Department at (860) 343-5977.

We want you to find your dream home from our wide selection of home plans. However, if for some reason you find that the plan you have purchased from us does not meet your needs, then you may exchange that plan for any other plan in our collection. We allow you sixty days from your original invoice date to make an exchange. At the time of the exchange you will be charged a processing fee of 15% of the total amount of your original order plus the difference in price between the plans (if applicable) plus the cost to ship the new plans to you. Call our Customer Service Department at (860) 343-5977 for more information. Please Note: Reproducible vellums can only be exchanged if they are unopened.

Important Shipping Information

Please refer to the shipping charts on the order form for service availability for your specific plan number. Our delivery service must have a street address or Rural Route Box number — never a post office box. (PLEASE NOTE: Supplying a P.O. Box number <u>only</u> will delay the shipping of your order.) Use a work address if no one is home during the day.

Orders being shipped to APO or FPO must go via First Class Mail. Please include the proper postage.

For our International Customers, only Certified bank checks and money orders are accepted and must be payable in U.S. currency. For speed, we ship international orders Air Parcel Post. Please refer to the chart for the correct shipping cost.

Important Canadian Shipping Information

To our friends in Canada, we have a plan design affiliate in Kitchener, Ontario. This relationship will help you avoid the delays and charges associated with shipments from the United States. Moreover, our affiliate is familiar with the building requirements in your community and country. We prefer payments in U.S. Currency. If you, however, are sending Canadian funds please add 40% to the prices of the plans and shipping fees.

An Important Note About Building Code Requirements:

All plans are drawn to conform to one or more of the industry's major national building standards. However, due to the variety of local building regulations, your plan may need to be modified to comply with local requirements — snow loads, energy loads, seismic zones, etc. Do check them fully and consult your local building officials.

A few states require that all building plans used be drawn by an architect registered in that state. While having your plans reviewed and stamped by such an architect may be prudent, laws requiring non-conforming plans like ours to be completely redrawn forces you to unnecessarily pay very large fees. If your state has such a law, we strongly recommend you contact your state representative to protest.

The rendering, floor plans, and technical information contained within this publication are not guaranteed to be totally accurate. Consequently, no information from this publication should be used either as a guide to constructing a home or for estimating the cost of building a home. Complete blueprints must be purchased for such purposes.

Garlinghouse 1998 Blueprint Price Code Schedule

Additional sets with original order $30

PRICE CODE	A	B	C	D	E	F	G	H
8 SETS OF SAME PLAN	$375	$415	$455	$495	$535	$575	$615	$655
4 SETS OF SAME PLAN	$325	$365	$405	$445	$485	$525	$565	$605
1 SINGLE SET OF PLANS	$275	$315	$355	$395	$435	$475	$515	$555
VELLUMS	$485	$530	$575	$620	$665	$710	$755	$800
MATERIALS LIST	$40	$40	$45	$45	$50	$50	$55	$55
ITEMIZED ZIP QUOTE	$75	$80	$85	$85	$90	$90	$95	$95

Shipping — (Plans 1-84999)

	1-3 Sets	4-6 Sets	7+ & Vellums
Standard Delivery (UPS 2-Day)	$15.00	$20.00	$25.00
Overnight Delivery	$30.00	$35.00	$40.00

Shipping — (Plans 85000-99999)

	1-3 Sets	4-6 Sets	7+ & Vellums
Ground Delivery (7-10 Days)	$9.00	$18.00	$20.00
Express Delivery (3-5 Days)	$15.00	$20.00	$25.00

International Shipping & Handling

	1-3 Sets	4-6 Sets	7+ & Vellums
Regular Delivery Canada (7-10 Days)	$14.00	$17.00	$20.00
Express Delivery Canada (5-6 Days)	$35.00	$40.00	$45.00
Overseas Delivery Airmail (2-3 Weeks)	$45.00	$52.00	$60.00

Option Key

- Zip Quote Available
- Duplex Plan
- Right Reading Reverse
- Materials List Available

Index

TOP SELLING
GARAGE PLANS

Save money by Doing-It-Yourself using our Easy-To-Follow plans. Whether you intend to build your own garage or contract it out to a building professional, the Garlinghouse garage plans provide you with everything you need to price out your project and get started. Put our 90+ years of experience to work for you. *Order now!!*

No. 06016C $86.00
Apartment Garage With One Bedroom

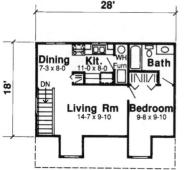

- 24' x 28' Overall Dimensions
- 544 Square Foot Apartment
- 12/12 Gable Roof with Dormers
- Slab or Stem Wall Foundation Options

No. 06015C $86.00
Apartment Garage With Two Bedrooms

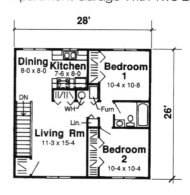

- 26' x 28' Overall Dimensions
- 728 Square Foot Apartment
- 4/12 Pitch Gable Roof
- Slab or Stem Wall Foundation Options

No. 06012C $54.00
30' Deep Gable &/or Eave Jumbo Garages

- 4/12 Pitch Gable Roof
- Available Options for Extra Tall Walls, Garage & Personnel Doors, Foundation, Window, & Sidings
- Package contains 4 Different Sizes
- 30' x 28'   • 30' x 32'   • 30' x 36'   • 30' x 40'

No. 06013C $68.00
Two-Car Garage With Mudroom/Breezeway

- Attaches to Any House
- 24' x 24' Eave Entry
- Available Options for Utility Room with Bath, Mudroom, Screened-In Breezeway, Roof, Foundation, Garage & Personnel Doors, Window, & Sidings

No. 06001C $48.00
12', 14' & 16' Wide-Gable 1-Car Garages

- Available Options for Roof, Foundation, Window, Door, & Sidings
- Package contains 8 Different Sizes
- 12' x 20' Mini-Garage • 14' x 22' • 16' x 20' • 16' x 24'
- 14' x 20'   • 14' x 24' • 16' x 22' • 16' x 26'

No. 06003C $48.00
24' Wide-Gable 2-Car Garages

- Available Options for Side Shed, Roof, Foundation, Garage & Personnel Doors, Window, & Sidings
- Package contains 5 Different Sizes
- 24' x 22' • 24' x 24' • 24' x 26'
- 24' x 28' • 24' x 32'

No. 06007C $60.00
Gable 2-Car Gambrel Roof Garages

- Interior Rear Stairs to Loft Workshop
- Front Loft Cargo Door With Pulley Lift
- Available Options for Foundation, Garage & Personnel Doors, Window, & Sidings
- Package contains 5 Different Sizes
- 22' x 26' • 22' x 28' • 24' x 28' • 24' x 30' • 24' x 32'

No. 06006C $48.00
22' & 24' Deep Eave 2 & 3-Car Garages

- Can Be Built Stand-Alone or Attached to House
- Available Options for Roof, Foundation, Garage & Personnel Doors, Window, & Sidings
- Package contains 6 Different Sizes
- 22' x 28' • 22' x 32' • 24' x 32'
- 22' x 30' • 24' x 30' • 24' x 36'

No. 06002C $48.00
20' & 22' Wide-Gable 2-Car Garages

- Available Options for Roof, Foundation, Garage & Personnel Doors, Window, & Sidings
- Package contains 7 Different Sizes
- 20' x 20' • 20' x 24' • 22' x 22' • 22' x 28'
- 20' x 22' • 20' x 28' • 22' x 24'

No. 06008C $60.00
Eave 2 & 3-Car Clerestory Roof Garages

- Interior Side Stairs to Loft Workshop
- Available Options for Engine Lift, Foundation, Garage & Personnel Doors, Window, & Sidings
- Package contains 4 Different Sizes
- 24' x 26' • 24' x 28' • 24' x 32' • 24' x 36'

Here's What You Get

- Three complete sets of drawings for each plan ordered

- Detailed step-by-step instructions with easy-to-follow diagrams on how to build your garage (not available with apartment garages)

- For each garage style, a variety of size and garage door configuration options

- Variety of roof styles and/or pitch options for most garages

- Complete materials list

- Choice between three foundation options: Monolithic Slab, Concrete Stem Wall or Concrete Block Stem Wall

- Full framing plans, elevations and cross-sectionals for each garage size and configuration

Build-It-Yourself PROJECT PLAN

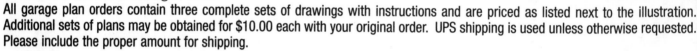

Order Information For Garage Plans:

All garage plan orders contain three complete sets of drawings with instructions and are priced as listed next to the illustration. Additional sets of plans may be obtained for $10.00 each with your original order. UPS shipping is used unless otherwise requested. Please include the proper amount for shipping.

Garage Order Form

Order Code No.   G8CH3

Please send me 3 complete sets of the following **GARAGE PLAN:**

Item no. & description	Price
_____	$ _____

Additional Sets

_____ (@ $10.00 each) $ _____

Shipping Charges: UPS-$3.75, First Class- $4.50 $ _____

Subtotal: $ _____

Resident sales tax: KS-6.15%, CT-6% $ _____
(NOT REQUIRED FOR OTHER STATES)

Total Enclosed: $ _____

Send your order to:
(With check or money order payable in U.S. funds only)
The Garlinghouse Company
P.O. Box 1717
Middletown, CT 06457

No C.O.D. orders accepted; U.S. funds only. UPS will not ship to Post Office boxes, FPO boxes, APO boxes, Alaska or Hawaii. Canadian orders must be shipped First Class.

Prices subject to change without notice.

My Billing Address is:

Name _____

Address _____

City _____

State _____ Zip_____

Daytime Phone No. _____

My Shipping Address is:

Name _____

Address _____
(UPS will not ship to P.O. Boxes)

City _____

State _____ Zip _____

For Faster Service...Charge It!
U.S. & Canada Call
1(800)235-5700
All foreign residents call 1(860)343-5977

❏ Mastercard ❏ Visa

Card # | | | | | | | | | | | | | | | | |

Signature _____ Exp.____/____

If paying by credit card, to avoid delays:
billing address must be as it appears on credit card statement

or FAX us at (860) 343-5984